Numerical Techniques in MATLAB®

In this book, various numerical methods are discussed in a comprehensive way. It delivers a mixture of theory, examples and MATLAB® practicing exercises to help the students in improving their skills. To understand the MATLAB programming in a friendly style, the examples are solved. The MATLAB codes are mentioned at the end of each topic. Throughout the text, a balance between theory, examples and programming is maintained.

Key Features

- Methods are explained with examples and codes.
- System of equations is given full consideration.
- Use of MATLAB is learned for every method.

This book is suitable for graduate students in mathematics, computer science and engineering.

Numerical Techniques in MATLAB®

Fundamental to Advanced Concepts

Dr. Taimoor Salahuddin

CRC Press
Taylor & Francis Group
Boca Raton London New York

CRC Press is an imprint of the
Taylor & Francis Group, an **informa** business

A CHAPMAN & HALL BOOK

First edition published 2024
by CRC Press
6000 Broken Sound Parkway NW, Suite 300, Boca Raton, FL 33487-2742

and by CRC Press
4 Park Square, Milton Park, Abingdon, Oxon, OX14 4RN

CRC Press is an imprint of Taylor & Francis Group, LLC

© 2024 Taimoor Salahuddin

ISBN: 9781032472584 (hbk)
ISBN: 9781032472577 (pbk)
ISBN: 9781003385288 (ebk)

DOI: 10.1201/9781003385288

Typeset in Times
by Deanta Global Publishing Services, Chennai, India

Contents

Author Bio...ix
Preface...xi

Chapter 1   Common Commands Used in MATLAB 1

    1.1   Basic Commands... 1
    1.2   Matrix Manipulation and Basic Notations 1
    1.3   Developing Arrays.. 3
    1.4   Cell Arrays 3
    1.5   Colon Operator 4
    1.6   Linspace... 4
    1.7   Ones... 4
    1.8   Zeros.. 4
    1.9   Eye 5
    1.10   Rand... 5
    1.11   Array Functions... 5
            1.11.1 Size... 5
            1.11.2 Length 6
            1.11.3 Dot.. 6
            1.11.4 Meshgrid... 6
            1.11.5 Reshape 6
            1.11.6 Det.. 7
    1.12   Two- and Three-Dimensional Plotting.. 7
            1.12.1 Hold On 7
            1.12.2 Grid On.. 7
            1.12.3 Two-Dimensional Graph 7
            1.12.4 Three-Dimensional Graph 8
    1.13   M.file Usage... 8
            1.13.1 Functions 8
    1.14   Use of Conditions (if, elseif, else).. 9
    1.15   Loop... 10
            1.15.1 For 10
    1.16   Break.. 10
    1.17   Return 10
    1.18   @ Command... 11
    1.19   Feval 11
    1.20   Inline Functions... 11
    1.21   Input... 12
    1.22   Fprintf.. 12

Chapter 2   System of Linear Equations 15

    2.1   Cramer's Rule 15
    2.2   Gauss Elimination Method.. 18
    2.3   Gauss-Jordan Elimination Method.. 20
    2.4   LU Factorization Methods.. 22
            2.4.1   Doolittle's Factorization Method ($L_{ii} = 1$) 22

2.4.2   Crout's Factorization Method ($U_{ii} = 1$) 28
2.4.3   Choleski's Factorization Method 31
2.5   Banded Coefficient Matrices 33
2.5.1   Doolittle's Factorization Method for Banded Tridiagonal Matrix 33
2.5.2   Crout's Factorization Method for Banded Tridiagonal Matrix 37
2.6   Banded Block Tridiagonal Matrices 41
2.6.1   Doolittle's Factorization Method for Banded Block Tridiagonal
Matrix 42
2.6.2   Crout's Factorization Method for Banded Block Tridiagonal
Matrix 46
2.7   Iterative Methods 50
2.7.1   Gauss-Jacobi Method 51
2.7.2   Gauss-Seidel Method 54
2.7.3   Conjugate Gradient Method 56
2.7.4   Convergence 60

Chapter 3   Polynomial Interpolation 65

3.1   Errors in Polynomial Interpolation 66
3.2   Newton's Forward Interpolation Formula (for Equal Intervals) 67
3.3   Newton's Backward Interpolation Formula (for Equal Intervals) 70
3.4   Newton's Divided Difference Interpolation (Unequal Intervals) 72
3.5   Lagrange's Interpolation Formula 75
3.6   Neville's Method 78
3.7   Cubic Spline Interpolation 80

Chapter 4   Root Finding Methods 89

4.1   Bisection Method 90
4.2   *Regula Falsi* Method 92
4.3   Newton Raphson Method 95
4.4   Secant Method 99
4.5   Newton Raphson Method for the System of Equations 101

Chapter 5   Numerical Integration 107

5.1   Newton-Cotes Formulas 107
5.1.1   The Trapezoidal Rule 108
5.1.2   Simpson's 1/3 Rule 111
5.1.3   Simpson's 3/8 Rule 115
5.2   Richardson Extrapolation 117
5.2.1   Romberg Integration 118
5.3   Adaptive Quadrature 122
5.4   Gaussian Quadrature 124

Chapter 6   Solution of Initial Value Problems (IVPs) 131

6.1   Single Step Methods 132
6.1.1   Euler's Method 132
6.1.1.1   Error Analysis 133
6.1.1.2   Euler's Method for the Systems of Ordinary
Differential Equations 135

6.1.2   Heun's Method 138
         6.1.2.1   Heun's Method for the Systems of Ordinary
                   Differential Equations 139
6.1.3   Modified Euler's Method.. 141
         6.1.3.1   Modified Euler's Method for the Systems of Ordinary
                   Differential Equations 143
6.1.4   Runge-Kutta Methods 144
         6.1.4.1   Second Order Runge-Kutta Methods 145
         6.1.4.2   Classical Fourth Order Runge-Kutta Method.................. 146
         6.1.4.3   Fourth Order Runge-Kutta Method for the Systems of
                   Ordinary Differential Equations.................................... 148
         6.1.4.4   Runge-Kutta-Fehlberg Technique.................................. 151
         6.1.4.5   Runge-Kutta-Fehlberg Technique for the Systems of
                   Ordinary Differential Equations.................................... 155
6.1.5   Multistep Methods... 160
         6.1.5.1   Adams Multistep Methods... 161
         6.1.5.2   Predictor-Corrector Methods by using Adams Formulas 163
         6.1.5.3   Predictor-Corrector Method for the Systems of
                   Ordinary Differential Equations.................................... 165
         6.1.5.4   Milne's Method... 169

Chapter 7 Boundary Value Problems (BVPs)... 175

7.1   Linear Shooting Method.. 175
7.2   Shooting Method for the Linear Ordinary Differential Systems 178
7.3   Shooting Method for Nonlinear Ordinary Differential Equations............... 183
      7.3.1   Secant Method Combined with Euler's Method: For Second
               Order Nonlinear Problems 183
      7.3.2   Secant Method Combined with Euler's Method: For Coupled
               Nonlinear Ordinary Differential Equations 188
      7.3.3   Newton Raphson Method Combined with Euler's Method: For
               Second Order Nonlinear Ordinary Differential Equations............. 193
      7.3.4   Newton Raphson Method Combined with Euler's Method: For
               Coupled Nonlinear Ordinary Differential Equations 197
7.4   Finite Difference Method for Linear Ordinary Differential Equations 202
      7.4.1   Finite Difference Approximation 202
      7.4.2   Central Difference Approximations..202
      7.4.3   Finite Difference Method for Second Order Linear Ordinary
               Differential Equations 203
      7.4.4   Finite Difference Method for the System of Linear Ordinary
               Differential Equations 207
7.5   Finite Difference Method for the Nonlinear Ordinary Differential
      Equations 211
      7.5.1   Finitee Difference Method for the Second Order Nonlinear
               Ordinary Differential Equations 211
      7.5.2   Finite Difference Method for the System of Nonlinear Ordinary
               Differential Equations 215

Appendix...225

References 247

Index 251

Author Bio

Dr Taimoor Salahuddin teaches in the department of Mathematics, at Mirpur University of Science and Technology (MUST), Mirpur AJK. He received a Ph.D. degree from QAU (Quaid-i-Azam), Islamabad in 2016. His journal research interests focus on numerical methods and fluid dynamics. After completing his Ph.D. degree, he joined MUST university as an Assistant Professor. He has more than 150 publications in impact factor journals. He received the highest citation award from the university in 2020. His love affair with computing began when his supervisor (Prof. Dr Malik Muhammad Yousaf) gave him an assignment to implement some problems on MATLAB®. Since then, he has published much work related to MATLAB.

Preface

This book has been designed specifically for research scholars who want to study and apply numerical methods in science and engineering by using MATLAB®. The language is comprehensive and sufficient explanation is provided so that scholars can get insight into the numerical techniques and understand them easily. I have been teaching this subject for the last six years and have witnessed that many students are facing problems understanding MATLAB. By keeping this in mind, I have written this book so that any science or engineering scholar can easily solve problems through MATLAB. The numerical methods are explained; however, the main theme is to objectify use of numerical methods through MATLAB software. Every presented method has been explained with examples. The main steps which are kept in mind for MATLAB programming are elaborated in the form of an algorithm and the codes are developed. Keeping in mind the average scholar, the writing technique is kept as simple as possible. The good thing about this book is that the programs are illustrated according to the solved examples. For better understanding, the examples are solved in detail. So every topic is divided into four parts: the method, the hand computations that examine the inner workings, the basic steps for solving the example in MATLAB and the program code.

The material consists of topics that are covered by engineers and science scholars in research for numerical purposes like the solution of equations, interpolation, root finding methods, numerical integration, and initial and boundary value problems.

Moreover, the system of equations has been fully emphasized. For example, in shooting and finite difference methods, the system of equations was less focused in the past. In this book the respected methods are explained in detail for the system of equations. This will aid in better understanding of the numerical execution of problems in research.

The material of this book is enough to cover a complete three credit hours course in addition to research. The programs mentioned in this book have been tested in MATLAB 2021a.

MATLAB is a registered trademark of The MathWorks, Inc. For product information, please contact:

The MathWorks, Inc.
3 Apple Drive
Natick, MA 01760-2098 USA
Tel: 508 647 7000
Fax: 508-647-7001
E-mail: info@mathworks.com
Web: www.mathworks.com

1 Common Commands Used in MATLAB®

MATLAB® is a very high-level programming language with several built-in packages that make the learner's interest in learning numerical techniques much easier. Nowadays, it is one of the major courses at higher educational level and research institutions. Numerical computations and visualization are easily performed in MATLAB. This language is a high-level array language with data structures, functions, control flow statements and object-oriented programming characteristics. It is very popular among students and teachers due to its three major advantages. Firstly, the algorithms can easily be constructed in different forms, so it allows the students to experiment with many numerical problems. Secondly, two-dimensional and three-dimensional graphs are easily illustrated within a script or in a command window. Thirdly, the most important advantage of MATLAB is its speed of calculating the data.

The main objective of this book is to deliver students a step-by-step procedure of the numerical methods, examples and their algorithms. In this chapter basic commands and operations will be introduced which will be helpful for learning the software and building a program for problem-solving.

1.1 BASIC COMMANDS

- In MATLAB, the command window can act as an electronic calculator in which each entry can be executed by pressing enter. Before starting examples, a few symbols will be kept in mind.
- The sign (>>) in MATLAB stands for quick input.
- A semicolon (;) has two advantages: (i) it hides the printout of results, (ii) it separates rows of the matrices.
- The sign (%) is used as the beginning of the comment.
- The sign 'ans' is referred to as the default name for outcome.
- Infinity is denoted by 'inf'.
- 'NaN' stands for not a number.
- The symbol 'π' is denoted by pi.
- Imaginary part is denoted by 'j' or 'i'.
- The smallest positive number is denoted by 'realmin'.
- The largest positive number is denoted by 'realmax'.
- The transpose of a matrix is denoted by '.
- A column is represented by ':'.
- The symbols -, +, *, ^, / and \ are referred to as subtraction, addition, multiplication, exponential, right division and left division.
- In matrices, the division, exponential and multiplication are added with one more symbol that is '.' i.e., .*, ./.
- The comparison operators <, >, <=, >=, == and ~= stand for less than, greater than, less than or equal to, greater than or equal to, equal to and not equal to.
- Few logical operators AND, OR and NOT are represented by &, | and ~.

1.2 MATRIX MANIPULATION AND BASIC NOTATIONS

Now by double-clicking the MATLAB sign on the desktop or searching MATLAB, the command window will open. By keeping these points in mind, a few examples in terms of matrices are given below:

DOI: 10.1201/9781003385288-1

```
>> % Command Window
>> A=[1 2 3;4 5 6;1 3 6]% A 3×3 matrix in which the printout result
%is shown
A =
   1   2   3
   4   5   6
   1   3   6


>> % Command Window
>> A=[1 2 3;4 5 6;1 3 6]; % The result hides by
%implementing the semicolon
>> B=[1;3;6]  % A column vector is achieved by inserting a semicolon
B =
   1
   3
   6

>> 10/0  % Infinity notation
ans =
   Inf
>> 0/0  % NaN notation
ans =
   NaN
>> pi % π notation
ans =
   3.1416
>> i % Imaginary number
ans =
   0.0000 + 1.0000i
>> eps    % Small number
ans =
   2.2204e-16
>> realmin      % Smallest positive number
ans =
 2.2251e-308
>> realmax   % Largest positive number
ans =
 1.7977e+308
>> A=[1 2 5; 8 5 2;-8 4 1];  % Transpose of a matrix
>> A'
ans =
   1   8  -8
   2   5   4
   5   2   1
>> A = 2 4 6;6 7 4;1 4 6]  % Consider a 3×3 matrix
A =
   2   4   6
   6   7   4
   1   4   6
>> A(:,2) % Represents the second column
ans =
   4
   7
   4
```

```
>> A=[3 4 6;3 5 5;6 8 5];
>> B=[3 3 5;5 7 3;9 7 4];
>> A+B    % Addition of matrices
ans =
    6    7   11
    8   12    8
   15   15    9
>> A=[3 4 6;3 5 5;6 8 5];
>> B=[3 3 5;5 7 3;9 7 4];
>> A-B   % Subtraction of matrices
ans =
    0    1    1
   -2   -2    2
   -3    1    1
>> A=[8 9 6;3 5 5;6 8 5];
>> B=[3 8 4;5 9 3;4 7 4];
>> A.*B    % Matrix multiplication

ans =
   24   72   24
   15   45   15
   24   56   20
>> A=[9 7 9;8 6 5;9 7 3];
>> B=[3 9 4;8 9 6;4 9 7];
>> A>B
ans =
    1    0    1
    0    0    0
    1    0    0
>> A=[8 7 3;2 3 4;2 1 2];
>> B=[3 3 2;3 4 6;1 3 4];
>> (A>B)|(B>2)

ans =
    1    1    1
    1    1    1
    1    1    1
% In comparison operators 1 hold for true and 0 stands for false
```

1.3   DEVELOPING ARRAYS

Arrays are simply created by writing elements between brackets.

```
>> A=[1 2 3 4 5]
A =
    1    2    3    4    5
```

1.4   CELL ARRAYS

A sequence of arbitrary objects is known as cell arrays. Cell arrays are formed by bounded them between the braces {}.

```
>> % Command Window
>> A={[1 2 3], [4 2 3],[2 9 3]}
A =
 1×3 cell array
  {[1 2 3]}  {[4 2 3]}  {[2 9 3]}
>> A{2}
ans =
    4    2    3
>> A{3}
ans =
    2    9    3
```

1.5   COLON OPERATOR

Equally spaced elements are constructed by using colon operator.

```
x = first term: increment: last term
```

```
>> x=0:0.1:1
x =
      0  0.1000  0.2000  0.3000  0.4000  0.5000  0.6000  0.7000
  0.8000  0.9000  1.0000
```

1.6   LINSPACE

Another way of creating arrays is by using the linspace operator.

```
x = linpsace(first term, last term, total number of iterations needed)
```

1.7   ONES

The function which has m rows and n columns filled with ones.

```
x = ones(m,n)
```

```
>> x=ones(2,3)
x =
    1    1    1
    1    1    1
```

1.8   ZEROS

This function works in a similar manner but fills the rows and columns with zeros.

```
x = zeros(m,n)
```

```
>>x=zeros(3,3)
x =
     0    0    0
     0    0    0
     0    0    0
```

1.9   EYE

This function creates an n×n identity matrix.

```
x = eye(n)
```

```
>> x=eye(3)
x =
     1    0    0
     0    1    0
     0    0    1
```

1.10   RAND

This function fills the matrix with random values between 0 and 1.

```
x = rand(m,n)
```

```
>> x=rand(3,3)
x =
   0.8147   0.9134   0.2785
   0.9058   0.6324   0.5469
   0.1270   0.0975   0.9575
```

1.11   ARRAY FUNCTIONS

Some of the basic array functions are:

1.11.1   SIZE

This command is helpful in calculating the number of rows and columns in a matrix.

```
[m,n] = size(x)
```

```
>> A=[1 2 3;4 5 7;2 4 6];
>> [m,n]=size(A)
m =
   3
n =
   3
```

1.11.2 Length

The number of elements of a vector can be calculated by this command.

n = length(x)

```
>> x=0:0.1:1;
>> length(x)
ans =
   11
```

1.11.3 Dot

The dot product of two vectors x and y having the same length can be calculated by:

C = dot(x,y)

1.11.4 Meshgrid

This command is used to transform the domain into arrays which are helpful in constructing three-dimensional graphs.

[X,Y] = meshgrid(x,y)

```
>> x=0:0.1:0.5;
>> y=0:0.1:0.5;
>> [X,Y]=meshgrid(x,y)
X =
        0   0.1000   0.2000   0.3000   0.4000   0.5000
        0   0.1000   0.2000   0.3000   0.4000   0.5000
        0   0.1000   0.2000   0.3000   0.4000   0.5000
        0   0.1000   0.2000   0.3000   0.4000   0.5000
        0   0.1000   0.2000   0.3000   0.4000   0.5000
        0   0.1000   0.2000   0.3000   0.4000   0.5000
Y =
        0        0        0        0        0        0
   0.1000   0.1000   0.1000   0.1000   0.1000   0.1000
   0.2000   0.2000   0.2000   0.2000   0.2000   0.2000
   0.3000   0.3000   0.3000   0.3000   0.3000   0.3000
   0.4000   0.4000   0.4000   0.4000   0.4000   0.4000
   0.5000   0.5000   0.5000   0.5000   0.5000   0.5000
```

1.11.5 Reshape

This command is useful to rearrange the elements into a matrix.

z = reshape(y,m,n)

```
>> y=0:1:5
y =
   0   1   2   3   4   5
>> z=reshape(y,2,3)
z =
   0   2   4
   1   3   5
```

1.11.6 DET

The command det represents the determinant of a matrix.

```
    x = det(A)
```

```
>> A=[1 4 6;2 4 5;2 6 8];
>> det(A)
ans =
    2
```

1.12 TWO- AND THREE-DIMENSIONAL PLOTTING

MATLAB is one of the best softwares for plotting. Before starting to illustrate two-dimensional and three-dimensional graphs, a few more commands must be kept in mind which are:

1.12.1 HOLD ON

This command helps to permit overwriting of previous plots.

1.12.2 GRID ON

This will show a coordinate grid.

1.12.3 TWO-DIMENSIONAL GRAPH

Now moving towards the two-dimensional plotting, consider an example in the command window in which $y = ax^2 + 3x + 1$, $-10 \leq x \leq 10$ and $a = 1, 3$ and 5.

```
>> a=1;
>> x=-10:0.1:10; % Generate x-array
>> y=a*x.^2+3*x+1; % Generate y-array

>> plot(x,y)
>> hold on % Overwriting of existing plot
>> a=3;
>> y=a*x.^2+3*x+1;
>> plot(x,y)
>> hold on
>> a=5;
>> y=a*x.^2+3*x+1;
>> plot(x,y)
```

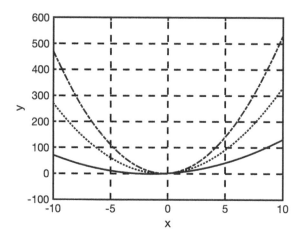

FIGURE 1.1 Two-dimensional graph.

1.12.4 THREE-DIMENSIONAL GRAPH

Consider an example in the command window in which $z = x^2 + y^2$, $-10 \leq x \leq 10$ and $-10 \leq y \leq 10$.

```
>> x=-10:0.1:10;    % Generate x-array
>> y=-10:0.1:10;    % Generate y-array
>> [X,Y]=meshgrid(x,y); % Transform the domain into arrays
>> Z=X.^2+Y.^2;     % Generate Z-array
>> surf(X,Y,Z)      % Three-dimensional plot
```

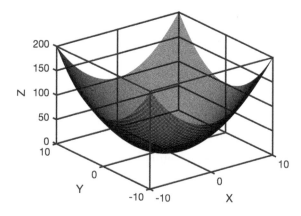

FIGURE 1.2 Three-dimensional graph.

1.13 M.FILE USAGE

When MATLAB opens, a sign of New Script will appear at the left top of the command window. By clicking that sign an untitled m.file will open. In order to understand how to use this file, a few commands and their usage will be kept in mind.

1.13.1 FUNCTIONS

In function, input and output arguments are defined which are separated by commas. The presentation of function is:

```
function [outputs] = name-of-function(inputs)
```

The function must be saved under the name-of-function.m. The results will be displayed in the command window. Moreover, the command window and m.file must be open in the same folder. Otherwise, results will not be displayed.

```
% M-file: file is saved by name 'abcd'
function [y]=abcd(x) % y is output and x is input
y=x.^2+9.*x+8;
end
```

```
% Command Window: copy [y]=abcd(x) and paste it in the command
% window
>> [y]=abcd(2)
y =
   30
```

1.14   USE OF CONDITIONS (IF, ELSEIF, ELSE)

These conditions are executed for true and false statements. First 'if' condition is implemented and it will block the statement if the condition is true. The condition will be skipped to 'elseif' if the condition is false, the block will be skipped and move forward to 'elseif'. Again, if the statement is false the block will be skipped and move forward to 'else' condition. The general algorithm is:

```
              If statement
                 block
          elseif statement
                 block
           else statement
                 block
                 end
```

```
% Mfile: Save the file by the name 'abcd'
function [abc]=abcd(x)
if x>10
  abc=1;
elseif x<10
  abc=-1;
else x=0
  abc=1
end
```

```
>> [abc]=abcd(15)
abc =
    1
```

```
% Copy [abc]=abcd(x)and paste in the Command Window
```

1.15 LOOP

1.15.1 FOR

This is used to target a sequence for which a target loop is achieved. The form is:

```
for target = sequence
block
End

>>% Command Window
>> for x=-2:2
y=x.^2
end
y =
   4
y =
   1
y =
   0
y =
   1
y =
   4
```

1.16 BREAK

A loop can be finished by using the break statement.

```
% Mfile
function [y]=abc(x)
for x=0:25
y=x.^2
if x<2
  break
end
end
```

```
% Command Window
>> abc
y =
   0
ans =
   0
```

1.17 RETURN

A function can be forced to exit by using the return command. In the next example a function y = x + (x + 1)/2 is assumed in which $dx = (x + 1)/2$ and a condition $|dx| < 1$ is implemented. The domain is taken from x = [−2, 2], it can be seen that when the loop (−2, −1,0,1,2) reaches 1, dx will be 1 which is against the condition. So, the return command is used here and it implements the desired result.

Note that if the return command is removed then the required result will not be obtained.

```
% Mfile y=x+(x+1)/2, abs((x+1)/2)<1, x=[-2,2]
function [y]=abc(x)
for x=-2:2
  dx=(x+1)./2;
y=x+dx
if abs(dx)<1
  return
end
end

% Command Window
>> abc
y =
  -2.5000
ans =
  -2.5000
```

1.18   @ COMMAND

This command is helpful in calling a function from another function.

1.19   FEVAL

A function passed to another function is handled by feval. This command will be extensively used. For this consider an example of function $y = x^3 + 8x^2 + 1$, with $-2 \leq x \leq 2$.

```
% Mfile 1
function [y]=abc1(func) % Output and inputs
for x=-2:2
  y=feval(func,x) % feval is used as a handled function
end
End

% Mfile 2
function [y]=abcd(x)
y=x.^3+8*x.^2+1;   % Given function
end

% Command Window
>> [y]=abc1(@abcd)
```

1.20   INLINE FUNCTIONS

In this command, the value of a given function is calculated for different values of independent variables. The algorithm is:

Name-of-function = inline('function', 'variable1', 'variable2', 'variable3')

```
% Command Window
>> abc=inline('x^3+9*y^2+5','x','y')
abc =
   Inline function:
   abc(x,y) = x^3+9*y^2+5
>> abc(2,1)
ans =
  22
```

1.21 INPUT

It illustrates a prompt and waits for input. If input is an expression, it is calculated and turned into a value. The general algorithm is:

Value = input('prompt')

```
>> x=input('Expession=')
Expession=[1 2 3;3 4 5;3 4 5]
x =
     1    2    3
     3    4    5
     3    4    5
```

1.22 FPRINTF

MATLAB shows numerical outcomes with four decimal places. It can be changed by using this command. The general form is:

fprintf('format', list-of-items)

The format contains floating point representation (%w.df), newline character (\n) and exponential (%w.de) representation. Where d represents the number of digits after decimal places and w is the width of the field.

```
% Mfile 1
function [y]=abc1(func) % Output and inputs
for x=-5:5
  y=feval(func,x); % feval is used as the handled function
  fprintf('%8.3f%11.6f\n', x,y)
end
end
```

```
% Mfile 2
function [y]=abcd(x)
y=x.^3+8*x.^2+1;    % Given function
end
```

```
% Command Window
>> [y] = abc1(@abcd)
-5.000    76.000000
-4.000    65.000000
-3.000    46.000000
-2.000    25.000000
-1.000     8.000000
 0.000     1.000000
 1.000    10.000000
 2.000    41.000000
 3.000   100.000000
 4.000   193.000000
 5.000   326.000000
```

2 System of Linear Equations

The main focus of this chapter is to calculate solutions for systems of linear algebraic equations by using MATLAB®. In most physical problems, the system of equations is very large, and takes much time to calculate their solutions. Consider the example of a truss in which a lightweight structure is built of triangular elements. The system defines the equilibrium of the vertical and horizontal forces on each node. A single triangular system is illustrated in Figure 2.1. Assume that to hold the structure into equilibrium the forces in each member of the truss be f_{12}, f_{13} and f_{23}. Suppose v_1 and v_2 are unknown supporting forces at nodes 1 and 2, N_1 is the unknown horizontal bracing force at node 1 and at node 3 the weight of the structure is expressed by M_3. Moreover, assume that the forces act in the positive direction. Hence the system representing the horizontal and vertical equilibrium at the three nodes become:

Node 1

$$v_1 + f_{13}\sin(\alpha) = 0,$$

$$N_1 + f_{12} + f_{23}\cos(\alpha) = 0,$$

Node 2

$$v_2 + f_{23}\sin(\beta) = 0,$$

$$-f_{12} - f_{23}\cos(\beta) = 0,$$

Node 3

$$-f_{13}\sin(\alpha) - f_{23}\sin(\beta) = M_3,$$

$$-f_{13}\cos(\alpha) + f_{23}\cos(\beta) = 0. \tag{2.1}$$

Direct and iterative approaches are used to get the numerical solutions of linear algebraic equations. The direct methods studied in this book are Cramer's rule, Gaussian elimination, Gauss-Jordan and LU decomposition. The iterative methods like Jacobi, Gauss-Seidel and conjugate gradient methods are briefly discussed. These methods along with examples and their usage in MATLAB are elaborated in detail.

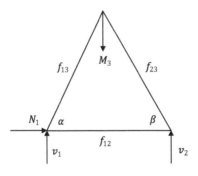

FIGURE 2.1 Graphical illustration of truss.

DOI: 10.1201/9781003385288-2

2.1 CRAMER'S RULE

This rule is the fraction of two determinants, the numerator is calculated by replacing the left column with the constant coefficients which appear on the right side of equations. The constants are replaced from left to right in order to calculate the unknowns.

Consider the system of linear equations in the following form:

$$a_{11}x_1 + a_{12}x_2 + a_{13}x_3 \dots a_{1n}x_n = b_1,$$
$$a_{21}x_1 + a_{22}x_2 + a_{23}x_3 \dots a_{2n}x_n = b_2,$$
$$- \qquad - \qquad - \qquad -$$
$$- \qquad - \qquad - \qquad - \qquad (2.2)$$
$$a_{n1}x_1 + a_{n2}x_2 + a_{n3}x_3 \dots a_{nn}x_n = b_n.$$

Where the constants and coefficients are known but x_n are unknown. In matrix form the Eq. (2.2) becomes

$$\begin{bmatrix} a_{11} & a_{12} & a_{13} & \dots & a_{1n} \\ a_{21} & a_{22} & a_{23} & \dots & a_{2n} \\ - & - & - & & - \\ - & - & - & & - \\ a_{n1} & a_{n2} & a_{n3} & \dots & a_{nn} \end{bmatrix} \begin{bmatrix} x_1 \\ x_2 \\ - \\ - \\ x_n \end{bmatrix} = \begin{bmatrix} b_1 \\ b_2 \\ - \\ - \\ b_n \end{bmatrix}, \qquad (2.3)$$

where

$$A = \begin{bmatrix} a_{11} & a_{12} & a_{13} & \dots & a_{1n} \\ a_{21} & a_{22} & a_{23} & \dots & a_{2n} \\ - & - & - & & - \\ - & - & - & & - \\ a_{n1} & a_{n2} & a_{n3} & \dots & a_{nn} \end{bmatrix}, X = \begin{bmatrix} x_1 \\ x_2 \\ - \\ - \\ x_n \end{bmatrix}, B = \begin{bmatrix} b_1 \\ b_2 \\ - \\ - \\ b_n \end{bmatrix}, \qquad (2.4)$$

determinant of A is

$$D = \begin{vmatrix} a_{11} & a_{12} & a_{13} & \dots & a_{1n} \\ a_{21} & a_{22} & a_{23} & \dots & a_{2n} \\ - & - & - & & - \\ - & - & - & & - \\ a_{n1} & a_{n2} & a_{n3} & \dots & a_{nn} \end{vmatrix} \neq 0, \qquad (2.5)$$

in this rule

$$X_n = \frac{|A_n|}{|A|}.$$

Where A_j's are calculated by replacing the j-th column of A by column vector B.

$$|A_1| = \begin{vmatrix} b_1 & a_{12} & a_{13} & \dots & a_{1n} \\ b_2 & a_{22} & a_{23} & \dots & a_{2n} \\ - & - & - & & - \\ - & - & - & & - \\ b_n & a_{n2} & a_{n3} & \dots & a_{nn} \end{vmatrix}, \qquad (2.6)$$

$$|A_2| = \begin{vmatrix} a_{11} & b_1 & a_{13} & \cdots\cdots & a_{1n} \\ a_{21} & b_2 & a_{23} & \cdots\cdots & a_{2n} \\ - & - & - & & - \\ - & - & - & & - \\ a_{n1} & b_n & a_{n3} & \cdots\cdots & a_{nn} \end{vmatrix},$$ (2.7)

$$\cdots\cdots\cdots$$

$$|A_n| = \begin{vmatrix} a_{11} & a_{12} & a_{13} & \cdots\cdots & b_1 \\ a_{21} & a_{22} & a_{23} & \cdots\cdots & b_2 \\ - & - & - & & - \\ - & - & - & & - \\ a_{n1} & a_{n2} & a_{n3} & \cdots\cdots & b_n \end{vmatrix}.$$ (2.8)

Example 2.1: Use Cramer's rule to calculate the solution:

$$x_1 + x_2 + x_3 = 3,$$

$$2x_1 + 3x_2 + 4x_3 = 2,$$

$$x_1 + 4x_2 + 9x_3 = 7.$$

Solution: The determinant D can be calculated as:

$$D = \begin{vmatrix} 3 & 4 \\ 4 & 9 \end{vmatrix} - \begin{vmatrix} 2 & 4 \\ 1 & 9 \end{vmatrix} + \begin{vmatrix} 2 & 3 \\ 1 & 4 \end{vmatrix} = 2,$$

substitution gives

$$x_1 = \frac{\begin{vmatrix} 3 & 1 & 1 \\ 2 & 3 & 4 \\ 7 & 4 & 9 \end{vmatrix}}{2} = 15,$$

$$x_2 = \frac{\begin{vmatrix} 1 & 3 & 1 \\ 2 & 2 & 4 \\ 1 & 7 & 9 \end{vmatrix}}{2} = -20,$$

$$x_3 = \frac{\begin{vmatrix} 1 & 1 & 3 \\ 2 & 3 & 2 \\ 1 & 4 & 7 \end{vmatrix}}{2} = 8.$$

Main steps for solving Cramer's rule for the systems of n equations in MATLAB: These steps will be kept in mind while implementing Cramer's rule:

1) The program is implemented in Mfile.
2) The program starts with a function command with output and inputs.
3) The length of b_1, b_2...b_n and coefficient matrix remains the same (n entries).
4) Define unknowns by n rows and one column, i.e.,

   X = zeros(n,1).
5) Store the matrix by some name.
6) A loop is implemented from 1:n to calculate the replaced column vectors and unknowns.
7) Each column vector is replaced with coefficients b_1, $b_2 \ldots b_n$.

8) Unknowns are calculated by using $X_n = \dfrac{|A_n|}{|A|}$.
9) Restore the matrix by the same name.

```
% Mfile
function [X]=cramer(A,b) % Inputs and Outputs
n=length(b); % The length of coefficient matrix
X=zeros(n,1); % Define unknowns by n-rows and 1 column
d=det(A);
abc=A; % Storing matrix by some name.
for j=1:n
  if d~=0 % The determinant should be non-zero
  A(:,j)=b;  % Each column vector is replaced
% with the coefficients b1, b2,…,bn
  X(j)=det(A)/d; % Unknowns are calculated
  A=abc;   % Restoring the matrix by same name
  end
end
end
```

```
% Command Window
>> [X]=cramer([1 1 1;2 3 4;1 4 9],[3 2 7])
X =
  15.0000
 -20.0000
   8.0000
```

2.2   GAUSS ELIMINATION METHOD

In this method, the n equations and n unknowns are reduced into the upper triangular matrix and then solved by back substitution. So this method involves two phases: elimination phase and back substitution phase [1].

Example 2.2: Consider the previous example.

$$x_1 + x_2 + x_3 = 3,$$

$$2x_1 + 3x_2 + 4x_3 = 2,$$

$$x_1 + 4x_2 + 9x_3 = 7.$$

Solution: The augmented coefficient matrix is

$$\begin{bmatrix} 1 & 1 & 1 & 3 \\ 2 & 3 & 4 & 2 \\ 1 & 4 & 9 & 7 \end{bmatrix}.$$

Here **1**, **3** and **9** are pivot elements. In the first step, the entries below the first pivot elements must be zero. For this row operation is performed, i.e.,

$$R_2 \leftarrow R_2 - \left(\frac{R_{21}}{R_{11}}\right) R_1,$$

$$R_3 \leftarrow R_3 - \left(\frac{R_{31}}{R_{11}}\right) R_1.$$

The matrix reduces to

$$\begin{bmatrix} 1 & 1 & 1 & 3 \\ 0 & 1 & 2 & -4 \\ 0 & 3 & 8 & 4 \end{bmatrix}.$$

New pivot elements are obtained. Again, performing the same procedure for the second pivot element which is **1**.

$$R_3 \leftarrow R_3 - \left(\frac{R_{32}}{R_{22}}\right) R_2.$$

The matrix reduces to

$$\begin{bmatrix} 1 & 1 & 1 & 3 \\ 0 & 1 & 2 & -4 \\ 0 & 0 & 2 & 16 \end{bmatrix}.$$

Now performing the back substitution procedure, we get

$$2x_3 = 16, x_2 + 2x_3 = -4, x_1 + x_2 + x_3 = 3,$$

the solution will be

$$x_1 = 15, x_2 = -20, x_3 = 8.$$

Main steps for solving the Gauss elimination method for the systems of n equations in MATLAB:
Consider an augmented matrix, where the constants are denoted by b:

1) The inputs and output are defined first. The solution X will be output and matrix A and constants are inputs.
2) As the matrix is $n \times n$, so constants b's must have length n.
3) The constants are column vectors.
4) The solution X has n rows and one column.
5) The first row is fixed and the entries below the first pivot elements should be zero. As in the augmented matrix there are $n - 1$ rows below pivot rows so j starts from 1:n − 1, and for transformed rows the loop starts from i = j + 1:n. So
   m = A(i,j)/A(j,j),
   A(i,:n) = A(i,j:n) − m * A(j,j:n).
      What will happen is that the elements below the first pivot entry will be zero and then the entries below the second pivot element will be zero and so on.
6) At the end back substitution is implemented.

```
% Mfile
function [X]=gauss_elimination(A,b) % Output X and inputs A and b
n=length(b); % Length must be same
b=b'; % Column vector
X=zeros(n,1); % X has n rows and one column
for j=1:n-1
  for i=j+1:n
    if A(i,j)~=0
    m=A(i,j)/A(j,j);
    A(i,j:n)=A(i,j:n)-m*A(j,j:n); % Transformed rows
      b(i)=b(i)-m*b(j);
    end
  end
end
for j=n:-1:1 % Back substitution
  X(j)=(b(j)-A(j,j+1:n)*X(j+1:n))/A(j,j);
end
```

```
>> % Command Window
>> [X]=gauss_elimination([1 1 1;2 3 4;1 4 9],[3 2 7])
X =
  15
 -20
   8
```

2.3 GAUSS-JORDAN ELIMINATION METHOD

In this method, a diagonal matrix is formed rather than an upper triangular matrix. This method is an extended form of Gauss elimination method. In Gauss elimination method, the elimination is done only in the lower elements of the pivots. But in the Gauss-Jordan elimination method, elimination is done both in lower and upper elements of pivots [1].

Example 2.3: Solve the previous example by using the Gauss-Jordan elimination method:

$$x_1 + x_2 + x_3 = 3,$$

$$2x_1 + 3x_2 + 4x_3 = 2,$$

$$x_1 + 4x_2 + 9x_3 = 7.$$

Solution: The augmented coefficient matrix form is

$$\begin{bmatrix} 1 & 1 & 1 & 3 \\ 2 & 3 & 4 & 2 \\ 1 & 4 & 9 & 7 \end{bmatrix}.$$

Here **1**, **3** and **9** are pivot elements. The entries below the first pivot elements must be zero. For this row operation is performed, i.e.,

$$R_2 \leftarrow R_2 - \left(\frac{R_{21}}{R_{11}}\right)R_1,$$

$$R_3 \leftarrow R_3 - \left(\frac{R_{31}}{R_{11}}\right) R_1.$$

The matrix reduces to

$$\begin{bmatrix} 1 & 1 & 1 & 3 \\ 0 & 1 & 2 & -4 \\ 0 & 3 & 8 & 4 \end{bmatrix}.$$

New pivot elements are obtained. Again, performing the same procedure for the second pivot element which is **1**.

$$R_3 \leftarrow R_3 - \left(\frac{R_{32}}{R_{22}}\right) R_2.$$

Note that in this method the entries above pivot elements should be zero. But in MATLAB when the first operation is finished then upper entries will be eliminated.

The matrix becomes

$$\begin{bmatrix} 1 & 1 & 1 & 3 \\ 0 & 1 & 2 & -4 \\ 0 & 0 & 2 & 16 \end{bmatrix}.$$

Now entries above **1** and **2** will be eliminated:

$$R_1 \leftarrow R_1 - \left(\frac{R_{12}}{R_{22}}\right) R_2,$$

$$R_2 \leftarrow R_2 - \left(\frac{R_{23}}{R_{33}}\right) R_3,$$

$$R_1 \leftarrow R_1 - \left(\frac{R_{13}}{R_{33}}\right) R_3.$$

Which gives

$$\begin{bmatrix} 1 & 0 & 0 & 15 \\ 0 & 1 & 0 & -20 \\ 0 & 0 & 2 & 16 \end{bmatrix}.$$

Hence we get

$$x_1 = 15, x_2 = -20, x_3 = 8.$$

Main steps for solving the Gauss-Jordan elimination method for the systems of n equations in MATLAB: Consider an augmented matrix:

1) The inputs and output are defined first. The solution X will be output and matrix A and constants are inputs.
2) As the matrix is $n \times n$, so constants b's must have length n.
3) The constants are column vectors.
4) X has n rows and one column.
5) The first row is fixed and the entries below the first pivot elements should be zero. As in the augmented matrix there are n − 1 rows below pivot rows so we implement j = 1:n − 1, for transformed rows the loop starts from i = j + 1:n.

6) Now for the diagonal matrix the elements above pivots will be eliminated. For this j will start from n to 1 with –1 increment (reverse process) and i starts from j – 1 to 1.
7) At the end back substitution is performed.

```
% Mfile
function [X]=gauss_jordan_elimination(A,b)  % Output X and inputs A
%and b
n=length(b); % Length must be same
b=b'; % Column vector
X=zeros(n,1); % X has n rows and one column
for j=1:n-1 % Pivot rows
  for i=j+1:n % Rows to be transformed
    if A(i,j)~=0
      m=A(i,j)/A(j,j);
    A(i,j:n)=A(i,j:n)-m*A(j,j:n); % Transformed rows
    b(i)=b(i)-m*b(j);
    end
  end
end
for j=n:-1:1  % For diagonal matrix
  for i=j-1:-1:1 % Rows to be transformed
    if A(i,j)~=0
      m=A(i,j)/A(j,j);
      A(i,j:n)=A(i,j:n)-m*A(j,j:n);
      b(i)=b(i)-m*b(j);
    end
  end
end
for j=n:-1:1  % Back substitution
  X(j)=(b(j)-A(j,j+1:n)*X(j+1:n))/A(j,j);
end
```

```
>>% Command Window
>> [X]=gauss_jordan_elimination([1 1 1;2 3 4;1 4 9],[3 2 7])
```

2.4   LU FACTORIZATION METHODS

The methods also known as decomposition methods composed of upper triangular matrix U and lower triangular matrix L. For any square matrix we have:

$$A = LU. \tag{2.9}$$

The way of calculating L and U for a square matrix is called LU factorization or LU decomposition. Doolittle's, Crout's and Choleski's are the most commonly used decomposition methods. Each method is explained with MATLAB code.

2.4.1   DOOLITTLE'S FACTORIZATION METHOD ($L_{ii}=1$)

Consider a 3×3 matrix having three unknowns. The system has the form:

$$a_{11}x_1 + a_{12}x_2 + a_{13}x_3 = b_1,$$

$$a_{21}x_1 + a_{22}x_2 + a_{23}x_3 = b_2,$$

$$a_{31}x_1 + a_{32}x_2 + a_{33}x_3 = b_3. \tag{2.10}$$

This system is equal to $AX = B$.

Where

$$A = \begin{bmatrix} a_{11} & a_{12} & a_{13} \\ a_{21} & a_{22} & a_{23} \\ a_{31} & a_{32} & a_{33} \end{bmatrix}, X = \begin{bmatrix} x_1 \\ x_2 \\ x_3 \end{bmatrix}, B = \begin{bmatrix} b_1 \\ b_2 \\ b_3 \end{bmatrix}. \tag{2.11}$$

The matrix A has the following form:

$$A = LU. \tag{2.12}$$

For Doolittle's method L and U are taken in the following form:

$$L = \begin{bmatrix} 1 & 0 & 0 \\ l_{21} & 1 & 0 \\ l_{31} & l_{32} & 1 \end{bmatrix}, U = \begin{bmatrix} u_{11} & u_{12} & u_{13} \\ 0 & u_{22} & u_{23} \\ 0 & 0 & u_{33} \end{bmatrix}. \tag{2.13}$$

After multiplication Eq. (2.12) becomes:

$$A = \begin{bmatrix} u_{11} & u_{12} & u_{13} \\ u_{11}l_{21} & u_{12}l_{21} + u_{22} & u_{13}l_{21} + u_{23} \\ u_{11}l_{31} & u_{12}l_{31} + u_{22}l_{32} & u_{13}l_{31} + u_{23}l_{32} + u_{33} \end{bmatrix}. \tag{2.14}$$

Where

$$A = \begin{bmatrix} a_{11} & a_{12} & a_{13} \\ a_{21} & a_{22} & a_{23} \\ a_{31} & a_{32} & a_{33} \end{bmatrix}. \tag{2.15}$$

Comparing Eq. (2.14) and Eq. (2.15)

$$\begin{bmatrix} a_{11} & a_{12} & a_{13} \\ a_{21} & a_{22} & a_{23} \\ a_{31} & a_{32} & a_{33} \end{bmatrix} = \begin{bmatrix} u_{11} & u_{12} & u_{13} \\ u_{11}l_{21} & u_{12}l_{21} + u_{22} & u_{13}l_{21} + u_{23} \\ u_{11}l_{31} & u_{12}l_{31} + u_{22}l_{32} & u_{13}l_{31} + u_{23}l_{32} + u_{33} \end{bmatrix}. \tag{2.16}$$

From Eq. (2.16) we get the L and U of the following form:

$$L = \begin{bmatrix} 1 & 0 & 0 \\ l_{21} = \dfrac{a_{21}}{u_{11}} & 1 & 0 \\ l_{31} = \dfrac{a_{31}}{u_{11}} & l_{32} = (a_{32} - u_{12}l_{31})/u_{22} & 1 \end{bmatrix}, \tag{2.17}$$

$$U = \begin{bmatrix} u_{11} = a_{11} & u_{12} = a_{12} & u_{13} = a_{13} \\ 0 & u_{22} = a_{22} - u_{12}l_{21} & u_{23} = a_{23} - u_{13}l_{21} \\ 0 & 0 & u_{33} = a_{33} - u_{13}l_{31} - u_{23}l_{32} \end{bmatrix}. \tag{2.18}$$

Put

$$LUX = B. \tag{2.19}$$

Assume

$$UX = Y, \tag{2.20}$$

and

$$LY = B. \tag{2.21}$$

First calculate $LY = B$ and then calculate $UX = Y$.
   Thus,

$$\begin{bmatrix} 1 & 0 & 0 \\ l_{21} & 1 & 0 \\ l_{31} & l_{32} & 1 \end{bmatrix} \begin{bmatrix} y_1 \\ y_2 \\ y_3 \end{bmatrix} = \begin{bmatrix} b_1 \\ b_2 \\ b_3 \end{bmatrix}. \tag{2.22}$$

Eq. (2.22) gives

$$y_1 = b_1,$$

$$l_{21}y_1 + y_2 = b_2,$$

$$l_{31}y_1 + l_{32}y_2 + y_3 = b_3. \tag{2.23}$$

By substitution, i.e., $UX = Y$

$$\begin{bmatrix} u_{11} & u_{12} & u_{13} \\ 0 & u_{22} & u_{23} \\ 0 & 0 & u_{33} \end{bmatrix} \begin{bmatrix} x_1 \\ x_2 \\ x_3 \end{bmatrix} = \begin{bmatrix} y_1 \\ y_2 \\ y_3 \end{bmatrix}. \tag{2.24}$$

Simplifying Eq. (2.24) gives

$$u_{11}x_1 + u_{12}x_2 + u_{13}x_3 = y_1,$$

$$u_{22}x_2 + u_{23}x_3 = y_2,$$

$$u_{33}x_3 = y_3. \tag{2.25}$$

This is the procedure for the Doolittle's factorization method.

Example 2.4: Consider

$$x_1 + x_2 + x_3 = 3,$$

$$2x_1 + 3x_2 + 4x_3 = 2,$$

$$x_1 + 4x_2 + 9x_3 = 7.$$

Find the solution of the problem by using Doolittle's factorization method.

Solution: Let

$$A = \begin{bmatrix} 1 & 1 & 1 \\ 2 & 3 & 4 \\ 1 & 4 & 9 \end{bmatrix}.$$

For Doolittle's method

$$L = \begin{bmatrix} 1 & 0 & 0 \\ l_{21} = \dfrac{a_{21}}{u_{11}} & 1 & 0 \\ l_{31} = \dfrac{a_{31}}{u_{11}} & l_{32} = \left(a_{32} - u_{12}l_{31}\right)/u_{22} & 1 \end{bmatrix},$$

$$U = \begin{bmatrix} u_{11} = a_{11} & u_{12} = a_{12} & u_{13} = a_{13} \\ 0 & u_{22} = a_{22} - u_{12}l_{21} & u_{23} = a_{23} - u_{13}l_{21} \\ 0 & 0 & u_{33} = a_{33} - u_{13}l_{31} - u_{23}l_{32} \end{bmatrix}.$$

From the above matrices we have $u_{11} = a_{11} = 1$, $u_{12} = a_{12} = 1$, $u_{13} = a_{13} = 1$, $l_{21} = \dfrac{a_{21}}{u_{11}} = \dfrac{2}{1} = 2$, $l_{31} = \dfrac{a_{31}}{u_{11}} = \dfrac{1}{1} = 1$, $u_{22} = a_{22} - u_{12}l_{21} = 3 - (1)(2) = 1$, $u_{23} = a_{23} - u_{13}l_{21} = 4 - (1)(2) = 2$, $l_{32} = (a_{32} - u_{12}l_{31})/u_{22} = (4 - (1)(1))/1 = 3$, $u_{33} = a_{33} - u_{13}l_{31} - u_{23}l_{32} = 9 - (1)(1) - (2)(3) = 2$.

After substituting the values in L and U, we have

$$U = \begin{bmatrix} 1 & 1 & 1 \\ 0 & 1 & 2 \\ 0 & 0 & 2 \end{bmatrix} \text{ and } L = \begin{bmatrix} 1 & 0 & 0 \\ 2 & 1 & 0 \\ 1 & 3 & 1 \end{bmatrix}.$$

Next, we take

$$\begin{bmatrix} 1 & 0 & 0 \\ l_{21} & 1 & 0 \\ l_{31} & l_{32} & 1 \end{bmatrix} \begin{bmatrix} y_1 \\ y_2 \\ y_3 \end{bmatrix} = \begin{bmatrix} b_1 \\ b_2 \\ b_3 \end{bmatrix}.$$

This gives

$$y_1 = b_1,$$

$$l_{21}y_1 + y_2 = b_2,$$

$$l_{31}y_1 + l_{32}y_2 + y_3 = b_3.$$

By putting the values of b_1, b_2, b_3, l_{21}, l_{31} and l_{32}, the values of y_1, y_2 and y_3 are calculated.
$y_1 = 3$, $y_2 = 2 - (2)(3) = -4$ and $y_3 = b_3 - l_{31}y_1 - l_{32}y_2 = 7 - (1)(3) - (3)(-4) = 16$.

Finally, by using the values of u_{11}, u_{12}, u_{13}, u_{22}, u_{23}, u_{33}, y_1, y_2 and y_3 we have

$$u_{11}x_1 + u_{12}x_2 + u_{13}x_3 = y_1,$$

$$u_{22}x_2 + u_{23}x_3 = y_2,$$

$$u_{33}x_3 = y_3.$$

Or

$$x_3 = \frac{y_3}{u_{33}} = 16/2 = 8, \qquad\qquad x_2 = (y_2 - u_{23}x_3)/u_{22} = (-4-(2)(8)/1 = -20,$$

$$x_1 = (y_1 - u_{12}x_2 - u_{13}x_3)/u_{11} = \big(3-(1)(-20)-(1)(8)\big)/(1) = 15.$$

Hence

$x_3 = 8, x_2 = -20$ and $x_1 = 15$.

Main steps for solving the Doolittle's factorization method for the systems of n equations in MATLAB:

1) The inputs will be A and b and X will be output.
2) The journal matrix will be n × n and the length of b will be n and in column form, i.e.,
   n = length(b),
   b = b′.
3) We write y's and X's in n rows and one column form, i.e.,
   X = zeros(n,1),
   y = zeros(n,1).
4) The upper and lower matrices will be n × n.
5) First, we take the upper diagonal entries. From the 3 × 3 matrix the upper diagonal entries are

   $$u_{11} = a_{11}, \ u_{22} = a_{22} - u_{12}l_{21}, \ u_{33} = a_{33} - u_{13}l_{31} - u_{23}l_{32}.$$

   From this relation we can see that this procedure is implemented in a loop from j = 1:n, the journal form in MATLAB can be written as:

   u (j,j) = a(j,j) – l(j,1:j – 1) * u(1:j – 1,j),

   for j = 1
   we have
   u(1,1) = a(1,1),
   for j = 2
   we have

   u(2,2) = a(2,2) – l(2,1)u(1,2),
   for j=3

   we have

   u(3,3) = a(3,3) – l(3,1)u(1,3 – l(3,2)u(2,3)

   and so on.

6) The diagonal entries for I can be written as
   l(j,j) = 1.
   As all entries are one.
7) The lower entries of I are

   $$l_{21} = \frac{a_{21}}{u_{11}}, \ l_{31} = \frac{a_{31}}{u_{11}}, \ l_{32} = (a_{32} - u_{12}l_{31})/u_{22}.$$

   The journal form is written on MATLAB by applying two loops which will start from

   j = 1:n, i = j + 1:n for l(i,j).
   What does this loop do? It will start from top to bottom after completing the first column (below pivot entry), it will move to the second column (below pivot entry) and so on.

   For example, for the 3 × 3 matrix, the loop is implemented in the following form:

   l(2,1),
   l(3,1).

After that it moves to the second column (below pivot entry) as

l(2,2),
l(3,2),
and finally
l(3,3).
In MATLAB, we can write it as
l(i,j) = (A(i,j) − l(i,1:j − 1) * u(1:j − 1,j))/u(j,j).

8) The upper matrix is also implemented in this loop because both l and u values are used at the same time, for u we can write it as
u(j,i) = A(j,i) − l(j,1:j − 1) * u(1:j − 1,i).
As can be seen, j appears first so the values achieved will be in row form as
u(1,2),
u(1,3).
Then the second row
u(2,3).

9) The values of y_1, y_2 and y_3 are calculated as the values are
$$y_1 = b_1, \; l_{21}y_1 + y_2 = b_2, \; l_{31}y_1 + l_{32}y_2 + y_3 = b_3.$$

In MATLAB the loop starts from 1 to n, the journal form in terms of y is
y(j) = b(j) − l(j,1:j − 1) * y(1:j − 1).

10) At the end back substitution is performed for the desired values of X.

```
% Mfile
function [X]=doolittle2(A,b) % Output X and input A and b
n=length(b); % Length must be same
b=b'; % Column vector
X=zeros(n,1); % X have n rows and one column
y=zeros(n,1); % y have n rows and one column
u=zeros(n,n);
l=zeros(n,n);
for j=1:n
  u(j,j)=A(j,j)-l(j,1:j-1)*u(1:j-1,j);% Diagonal matrix
  l(j,j)=1;
  for i=j+1:n
    l(i,j)=(A(i,j)-l(i,1:j-1)*u(1:j-1,j))/u(j,j);
% Lower diagonal matrix
    u(j,i)=A(j,i)-l(j,1:j-1)*u(1:j-1,i);
  % Upper diagonal matrix
   end
end
for j=1:n
  y(j)=b(j)-l(j,1:j-1)*y(1:j-1); % LY=B
end
for j=n:-1:1 % Back substitution UX=Y
  X(j)=(y(j)-u(j,j+1:n)*X(j+1:n))/u(j,j);
end


>> % Command Window
>> doolittle2([1 1 1;2 3 4;1 4 9],[3 2 7])
ans =
  15
 -20
   8
```

2.4.2 Crout's Factorization Method ($U_{ii}= 1$)

Consider a 3×3 matrix having three unknowns. The system becomes

$$a_{11}x_1 + a_{12}x_2 + a_{13}x_3 = b_1,$$

$$a_{21}x_1 + a_{22}x_2 + a_{23}x_3 = b_2,$$

$$a_{31}x_1 + a_{32}x_2 + a_{33}x_3 = b_3. \tag{2.26}$$

This system is equal to $AX = B$.
   Where

$$A = \begin{bmatrix} a_{11} & a_{12} & a_{13} \\ a_{21} & a_{22} & a_{23} \\ a_{31} & a_{32} & a_{33} \end{bmatrix}, X = \begin{bmatrix} x_1 \\ x_2 \\ x_3 \end{bmatrix}, B = \begin{bmatrix} b_1 \\ b_2 \\ b_3 \end{bmatrix}. \tag{2.27}$$

The matrix A has the following form:

$$A = LU.$$

   For Crout's method, L and U have the following form:

$$L = \begin{bmatrix} l_{11} & 0 & 0 \\ l_{21} & l_{22} & 0 \\ l_{31} & l_{32} & l_{33} \end{bmatrix}, U = \begin{bmatrix} 1 & u_{12} & u_{13} \\ 0 & 1 & u_{23} \\ 0 & 0 & 1 \end{bmatrix}. \tag{2.28}$$

After multiplication, we get

$$A = \begin{bmatrix} l_{11} & u_{12}l_{11} & u_{13}l_{11} \\ l_{21} & u_{12}l_{21} + l_{22} & u_{13}l_{21} + u_{23}l_{22} \\ l_{31} & u_{12}l_{31} + l_{32} & u_{13}l_{31} + u_{23}l_{32} + l_{33} \end{bmatrix}. \tag{2.29}$$

Where

$$A = \begin{bmatrix} a_{11} & a_{12} & a_{13} \\ a_{21} & a_{22} & a_{23} \\ a_{31} & a_{32} & a_{33} \end{bmatrix}. \tag{2.30}$$

Comparing Eqs. (2.29) and (2.30) gives

$$\begin{bmatrix} a_{11} & a_{12} & a_{13} \\ a_{21} & a_{22} & a_{23} \\ a_{31} & a_{32} & a_{33} \end{bmatrix} = \begin{bmatrix} l_{11} & u_{12}l_{11} & u_{13}l_{11} \\ l_{21} & u_{12}l_{21} + l_{22} & u_{13}l_{21} + u_{23}l_{22} \\ l_{31} & u_{12}l_{31} + l_{32} & u_{13}l_{31} + u_{23}l_{32} + l_{33} \end{bmatrix}. \tag{2.31}$$

The L and U have the following form

$$L = \begin{bmatrix} l_{11} = a_{11} & 0 & 0 \\ l_{21} = a_{21} & l_{22} = a_{22} - u_{12}l_{21} & 0 \\ l_{31} = a_{31} & l_{32} = \left(a_{32} - u_{12}l_{31} \right) & l_{33} = a_{33} - u_{13}l_{31} - u_{23}l_{32} \end{bmatrix}, \tag{2.32}$$

$$U = \begin{bmatrix} 1 & u_{12} = a_{12}/l_{11} & u_{13} = a_{13}/l_{11} \\ 0 & 1 & u_{23} = \left(a_{23} - u_{13}l_{21}\right)/l_{22} \\ 0 & 0 & 1 \end{bmatrix}. \tag{2.33}$$

Put

$$LUX = B.$$

The remaining procedure is the same as described for Doolittle's method.

Example 2.5: Consider

$$x_1 + x_2 + x_3 = 3,$$

$$2x_1 + 3x_2 + 4x_3 = 2,$$

$$x_1 + 4x_2 + 9x_3 = 7.$$

Solution: Let

$$A = \begin{bmatrix} 1 & 1 & 1 \\ 2 & 3 & 4 \\ 1 & 4 & 9 \end{bmatrix}.$$

For Crout's method

$$L = \begin{bmatrix} l_{11} = a_{11} & 0 & 0 \\ l_{21} = a_{21} & l_{22} = a_{22} - u_{12}l_{21} & 0 \\ l_{31} = a_{31} & l_{32} = \left(a_{32} - u_{12}l_{31}\right) & l_{33} = a_{33} - u_{13}l_{31} - u_{23}l_{32} \end{bmatrix},$$

$$U = \begin{bmatrix} 1 & u_{12} = a_{12}/l_{11} & u_{13} = a_{13}/l_{11} \\ 0 & 1 & u_{23} = \left(a_{23} - u_{13}l_{21}\right)/l_{22} \\ 0 & 0 & 1 \end{bmatrix}.$$

From the above matrices we have $l_{11} = a_{11} = 1$, $l_{21} = a_{21} = 2$, $l_{31} = a_{31} = 1$, $u_{12} = \dfrac{a_{12}}{l_{11}} = \dfrac{1}{1} = 1$,

$u_{13} = \dfrac{a_{13}}{l_{11}} = \dfrac{1}{1} = 1$, $l_{22} = a_{22} - u_{12}l_{21} = 3 - (1)(2) = 1$, $u_{23} = (a_{23} - u_{13}l_{21}) / l_{22} = \left(4 - (1)(2)\right)/1 = 2$,

$l_{32} = (a_{32} - u_{12}l_{31}) = \left(4 - (1)(1)\right) = 3$, $l_{33} = a_{33} - u_{13}l_{31} - u_{23}l_{32} = 9 - (1)(1) - (2)(3) = 2$.

So

$$U = \begin{bmatrix} 1 & 1 & 1 \\ 0 & 1 & 2 \\ 0 & 0 & 1 \end{bmatrix} \text{ and } L = \begin{bmatrix} 1 & 0 & 0 \\ 2 & 1 & 0 \\ 1 & 3 & 2 \end{bmatrix}.$$

Next

$$\begin{bmatrix} l_{11} & 0 & 0 \\ l_{21} & l_{22} & 0 \\ l_{31} & l_{32} & l_{33} \end{bmatrix} \begin{bmatrix} y_1 \\ y_2 \\ y_3 \end{bmatrix} = \begin{bmatrix} b_1 \\ b_2 \\ b_3 \end{bmatrix}.$$

This gives

$$l_{11}y_1 = b_1,$$

$$l_{21}y_1 + l_{22}y_2 = b_2,$$

$$l_{31}y_1 + l_{32}y_2 + l_{33}y_3 = b_3.$$

By putting the values of b_1, b_2 and b_3, l_{11}, l_{21}, l_{22}, l_{31}, l_{32} and l_{33} the values of y_1, y_2 and y_3 are calculated, i.e.,

$$y_1 = \frac{b_1}{l_{11}} = 3/1, \; y_2 = \frac{b_2 - l_{21}y_1}{l_{22}} = (2-(2)(3))/(1), \; y_3 = \frac{b_3 - l_{31}y_1 - l_{32}y_2}{l_{33}} = (7-(1)(3)-(3)(-4))/2.$$

Which gives $y_1 = 3$, $y_2 = -4$ and $y_3 = 8$.

Finally, by using Eq. (2.25) and putting the values of u_{11}, u_{12}, u_{13}, u_{22}, u_{23}, u_{33}, y_1, y_2 and y_3 it yields

$$x_1 = y_1 - u_{12}x_2 - u_{13}x_3,$$

$$x_2 = y_2 - u_{23}x_3,$$

$$x_3 = y_3.$$

By back substitution, we get

$x_3 = 8, x_2 = -4-(2)(8) = -20$ and $x_1 = 3-(1)(-20)-(1)(8) = 15$.

Main steps for solving the Crout's factorization method for the systems of n equations in MATLAB: In this method the lower and upper diagonal entries are changed. The remaining procedure is the same.

1) We start from lower diagonal entries which are

$l_{11} = a_{11}, l_{22} = a_{22} - u_{12}l_{21}, l_{33} = a_{33} - u_{13}l_{31} - u_{23}l_{32}$.

A loop from j = 1:n is implemented and from l_{33} we can journalize the relation, i.e.,

l(j,j) = a(j,j) −l(j,1:j − 1) * u(1:j − 1,j).

From above if we put j = 3 we get

l(3,3) = a(3,3) −l (3,1)u(1,3) − l(3,2)u(2,3).

From this we can see that l(j,1:j − 1) means that if we put j = 3 the first entry will be 3 but the second entry starts from 1 and then ends at 2 as we have j − 1.

2) The diagonal entries for u can be written as

u(j,j) = 1.

All entries are one.

3) The upper entries of u are

$$u_{21} = \frac{a_{21}}{l_{11}},$$

$$u_{31} = \frac{a_{31}}{l_{11}},$$

$$u_{32} = (a_{32} - u_{12}l_{31})/l_{22}.$$

And the lower entries of l are

$$l_{21} = a_{21},$$

$$l_{31} = a_{31},$$

$$l_{32} = (a_{32} - u_{12}l_{31}).$$

Both are implemented in the same loop as both values are used at the same time. From their third relation we can write the journal form as

$u(j,i) = (A(j,i) - l(j,1{:}j - 1) * u(1{:}j - 1,i))/l(j,j),$
$l(i,j) = A(i,j) - l(i,1{:}j - 1) * u(1{:}j - 1,j).$

The loop starts from j = 1:n and i = j + 1:n. For (i,j) it operates the first column as it starts from (2,1) and then the second column which starts from (3,2) and so on. For (j,i) it covers the rows in a similar fashion.

4) The values of y_1, y_2 and y_3 are calculated in a similar way as calculated for Doolittle's factorization method.

5) At the end back substitution is performed for desired values of X.

```
% Mfile
function [X]=Crouts(A,b)   % Output X and inputs A and b
n=length(b); % Length must be same
b=b'; % Column vector
X=zeros(n,1); % X have n rows and one column
y=zeros(n,1); % y have n rows and one column
u=zeros(n,n);
l=zeros(n,n);
for j=1:n  % Diagonal entries for l
   l(j,j)=A(j,j)-l(j,1:j-1)*u(1:j-1,j);
   u(j,j)=1;
   for i=j+1:n % Upper and lower entries for u and l
      u(j,i)=(A(j,i)-l(j,1:j-1)*u(1:j-1,i))/l(j,j);
      l(i,j)=A(i,j)-l(i,1:j-1)*u(1:j-1,j);
   end
end
for j=1:n
  y(j)=(b(j)-l(j,1:j-1)*y(1:j-1))/l(j,j); % LY=B
end
for j=n:-1:1  % Back substitution UX=Y
  X(j)=(y(j)-u(j,j+1:n)*X(j+1:n))/u(j,j);
end


>> % Command Window
>> Crouts([1 1 1;2 3 4;1 4 9],[3 2 7])
```

2.4.3 CHOLESKI'S FACTORIZATION METHOD

This method is based on a symmetric matrix which can be decomposed as

$$A = LL^T. \tag{2.34}$$

For this consider a 3 × 3 matrix, which can be decomposed as

$$A = \begin{bmatrix} a_{11} & a_{12} & a_{13} \\ a_{21} & a_{22} & a_{23} \\ a_{31} & a_{32} & a_{33} \end{bmatrix} = \begin{bmatrix} l_{11} & 0 & 0 \\ l_{21} & l_{22} & 0 \\ l_{31} & l_{32} & l_{33} \end{bmatrix} \begin{bmatrix} l_{11} & l_{12} & l_{13} \\ 0 & l_{22} & l_{23} \\ 0 & 0 & l_{33} \end{bmatrix}. \tag{2.35}$$

After multiplication we get

$$\begin{bmatrix} a_{11} & a_{12} & a_{13} \\ a_{21} & a_{22} & a_{23} \\ a_{31} & a_{32} & a_{33} \end{bmatrix} = \begin{bmatrix} l^2_{11} & l_{11}l_{12} & l_{11}l_{31} \\ l_{11}l_{12} & l^2_{21} + l^2_{22} & l_{21}l_{31} + l_{22}l_{32} \\ l_{11}l_{31} & l_{21}l_{31} + l_{22}l_{32} & l^2_{31} + l^2_{32} + l^2_{33} \end{bmatrix}. \tag{2.36}$$

In this matrix, there are six unknowns (total elements are nine but due to symmetric entries six will be left).

Note that a few entries in the matrix have a square root, so a condition will be imposed that entries must be positive otherwise the square root of negative entries will be omitted. Firstly, the unknown diagonal elements are

$$l_{11} = \sqrt{a_{11}}, l_{22} = \sqrt{a_{22} - l^2_{21}}, l_{33} = \sqrt{a_{33} - l^2_{31} - l^2_{32}}. \tag{2.37}$$

The off diagonal entries are

$$l_{21} - a_{21} / l_{11}, l_{31} = a_{31} / l_{11}, l_{32} = \left(a_{32} - l_{21}l_{31}\right) / l_{22}. \tag{2.38}$$

Example 2.6: Consider

$$A = \begin{bmatrix} 1 & 1 & 1 \\ 2 & 5 & 4 \\ 1 & 4 & 20 \end{bmatrix}.$$

Solution: Since

$$l_{11} = \sqrt{a_{11}}, \qquad l_{22} = \sqrt{a_{22} - l^2_{21}}, \qquad l_{33} = \sqrt{a_{33} - l^2_{31} - l^2_{32}}, \qquad l_{21} = a_{21} / l_{11}, \qquad l_{31} = a_{31} / l_{11},$$

$$l_{32} = \left(a_{32} - l_{21}l_{31}\right) / l_{22}.$$

Putting the values

$$l_{11} = \sqrt{1} = 1, \qquad l_{21} = \frac{a_{21}}{l_{11}} = \frac{2}{1} = 2, \qquad l_{22} = \sqrt{a_{22} - l^2_{21}} = \sqrt{5 - (2)^2} = 1, \qquad l_{31} = a_{31} / l_{11} = 1,$$

$$l_{32} = \frac{a_{32} - l_{21}l_{31}}{l_{22}} = \frac{4 - (2)(1)}{1} = 2, \quad l_{33} = \sqrt{20 - (1)^2 - (2)^2} = 3.8.$$

Main steps for solving Choleski's factorization method for the systems of n equations in MATLAB:

1) The input and output are defined first. The matrix L will be output and matrix A is input.
2) The journal matrix is $n \times n$.
3) The terms appearing in square root (diagonal elements) will be positive, for this a condition will be imposed.
4) The formula in the loop can be written as
   A(i,j) = (A(i,j) − dot(A(i,1:j − 1),A(j,1:j − 1)))/A(j,j), for j = 1 to n and i = j + 1 to n.
5) Finally, a command by the name of tril is implemented in order to get a lower triangular form.

```
% Mfile
function [L]=choleski_decomposition(A)
n=size(A)  % Square matrix
for j=1:n
  abc=A(j,j)-dot(A(j,1:j-1),A(j,1:j-1));
  if abc<0
    error ('matrix is negative')
  end
  A(j,j)=sqrt(abc);
  for i=j+1:n
    A(i,j)=(A(i,j)-dot(A(i,1:j-1),A(j,1:...
j-1)))/A(j,j);
  end
```

```
end
L=tril(A);
```

```
>> % Command Window
>> [L]=choleski_decomposition([1 1 1;2 5 4;1 4 20])
L =
   1.0000        0        0
   2.0000   1.0000        0
   1.0000   2.0000   3.8730
```

2.5   BANDED COEFFICIENT MATRICES

The non-zero terms assembled about the leading diagonal are said to be a banded coefficient matrix. For example:

$$A = \begin{bmatrix} X & X & 0 & 0 & 0 \\ X & X & X & 0 & 0 \\ 0 & X & X & X & 0 \\ 0 & 0 & X & X & X \\ 0 & 0 & 0 & X & X \end{bmatrix}. \tag{2.39}$$

In this matrix, X's denote the non-zero elements in which some of them may be zero. All the elements outside the band structure are zero. The given matrix is tridiagonal, it can be decomposed in the form A = LU, both L and U will maintain the banded structure of A. For example, if we decomposed the above matrix, we will get

$$L = \begin{bmatrix} X & 0 & 0 & 0 & 0 \\ X & X & 0 & 0 & 0 \\ 0 & X & X & 0 & 0 \\ 0 & 0 & X & X & 0 \\ 0 & 0 & 0 & X & X \end{bmatrix}, U = \begin{bmatrix} X & X & 0 & 0 & 0 \\ 0 & X & X & 0 & 0 \\ 0 & 0 & X & X & 0 \\ 0 & 0 & 0 & X & X \\ 0 & 0 & 0 & 0 & X \end{bmatrix}. \tag{2.40}$$

In this book, Doolittle's and Crout's methods are implemented to solve the following tridiagonal matrices.

2.5.1   DOOLITTLE'S FACTORIZATION METHOD FOR BANDED TRIDIAGONAL MATRIX

Consider a 3 × 3 tridiagonal matrix

$$A = \begin{bmatrix} a_1 & c_1 & 0 \\ b_2 & a_2 & c_2 \\ 0 & b_3 & a_3 \end{bmatrix}. \tag{2.41}$$

The matrix A can be written as

$$A = LU. \tag{2.42}$$

For Doolittle's method L and U are

$$L = \begin{bmatrix} 1 & 0 & 0 \\ \beta_2 & 1 & 0 \\ 0 & \beta_3 & 1 \end{bmatrix}, U = \begin{bmatrix} \alpha_1 & \gamma_1 & 0 \\ 0 & \alpha_2 & \gamma_2 \\ 0 & 0 & \alpha_3 \end{bmatrix}. \tag{2.43}$$

Multiplication of L and U gives

$$A = \begin{bmatrix} \alpha_1 & \gamma_1 & 0 \\ \beta_2\alpha_1 & \beta_2\gamma_1 + \alpha_2 & \gamma_2 \\ 0 & \beta_3\alpha_2 & \beta_3\gamma_2 + \alpha_3 \end{bmatrix}. \tag{2.44}$$

Comparing both A's

$$\begin{bmatrix} a_1 & c_1 & 0 \\ b_2 & a_2 & c_2 \\ 0 & b_3 & a_3 \end{bmatrix} = \begin{bmatrix} \alpha_1 & \gamma_1 & 0 \\ \beta_2\alpha_1 & \beta_2\gamma_1 + \alpha_2 & \gamma_2 \\ 0 & \beta_3\alpha_2 & \beta_3\gamma_2 + \alpha_3 \end{bmatrix}. \tag{2.45}$$

The diagonal entries are:

$$\alpha_1 = a_1, \quad \alpha_2 = a_2 - \beta_2\gamma_1, \quad \alpha_3 = a_3 - \beta_3\gamma_2. \tag{2.46}$$

The upper diagonal entries are:

$$\gamma_1 = c_1, \gamma_2 = c_2. \tag{2.47}$$

The lower diagonal entries are:

$$\beta_2 = \frac{b_2}{\alpha_1}, \beta_3 = \frac{b_3}{\alpha_2}. \tag{2.48}$$

We get L and U of the following form

$$L = \begin{bmatrix} 1 & 0 & 0 \\ \beta_2 = \dfrac{b_2}{\alpha_1} & 1 & 0 \\ 0 & \beta_3 = \dfrac{b_3}{\alpha_2} & 1 \end{bmatrix}, \tag{2.49}$$

$$U = \begin{bmatrix} \alpha_1 = a_1 & \gamma_1 = c_1 & 0 \\ 0 & \alpha_2 = a_2 - \beta_2\gamma_1 & \gamma_2 = c_2 \\ 0 & 0 & \alpha_3 = a_3 - \beta_3\gamma_2 \end{bmatrix}. \tag{2.50}$$

Put

$$LUX = R. \tag{2.51}$$

Calculate $LY = R$

$$\begin{bmatrix} 1 & 0 & 0 \\ \beta_2 & 1 & 0 \\ 0 & \beta_3 & 1 \end{bmatrix} \begin{bmatrix} y_1 \\ y_2 \\ y_3 \end{bmatrix} = \begin{bmatrix} r_1 \\ r_2 \\ r_3 \end{bmatrix}. \tag{2.52}$$

This gives

$$y_1 = r_1,$$

$$y_2 = r_2 - y_1 \beta_2 ,$$

$$y_3 = r_3 - y_2 \beta_3 . \tag{2.53}$$

Finally, calculate $UX = Y$

$$\begin{bmatrix} \alpha_1 & \gamma_1 & 0 \\ 0 & \alpha_2 & \gamma_2 \\ 0 & 0 & \alpha_3 \end{bmatrix} \begin{bmatrix} x_1 \\ x_2 \\ x_3 \end{bmatrix} = \begin{bmatrix} y_1 \\ y_2 \\ y_3 \end{bmatrix} . \tag{2.54}$$

Solving gives

$$x_3 = y_3 / \alpha_3,$$

$$x_2 = (y_2 - \gamma_2 x_3) / \alpha_2,$$

$$x_1 = (y_1 - \gamma_1 x_2) / \alpha_1. \tag{2.55}$$

Example 2.7: Solve the following tridiagonal system

$$\begin{bmatrix} 1 & 2 & 0 \\ 3 & 1 & 2 \\ 0 & 3 & 1 \end{bmatrix} \begin{bmatrix} x_1 \\ x_2 \\ x_3 \end{bmatrix} = \begin{bmatrix} 1 \\ 1 \\ 1 \end{bmatrix} .$$

Solution:
Let

$$A = \begin{bmatrix} 1 & 2 & 0 \\ 3 & 1 & 2 \\ 0 & 3 & 1 \end{bmatrix}, R = \begin{bmatrix} 1 \\ 1 \\ 1 \end{bmatrix} .$$

For Doolittle's tridiagonal matrix

$$\begin{bmatrix} 1 & 2 & 0 \\ 3 & 1 & 2 \\ 0 & 3 & 1 \end{bmatrix} = \begin{bmatrix} \alpha_1 & \gamma_1 & 0 \\ \beta_2 \alpha_1 & \beta_2 \gamma_1 + \alpha_2 & \gamma_2 \\ 0 & \beta_3 \alpha_2 & \beta_3 \gamma_2 + \alpha_3 \end{bmatrix} .$$

Comparing gives

$$\gamma_1 = 2, \gamma_2 = 2, \ \alpha_1 = 1, \ \beta_2 = 3, \ \alpha_2 = 1 - (3)(2) = -5, \ \beta_3 = -\frac{3}{5}, \ \alpha_3 = 1 - \left(\frac{-3}{5}\right)(2) = \frac{11}{5}.$$

So

$$U = \begin{bmatrix} 1 & 2 & 0 \\ 0 & -5 & 2 \\ 0 & 0 & 11/5 \end{bmatrix} \text{ and } L = \begin{bmatrix} 1 & 0 & 0 \\ 3 & 1 & 0 \\ 0 & -3/5 & 1 \end{bmatrix} .$$

Next

$$y_1 = 1,$$

$$y_2 = 1-(1)(3) = -2,$$

$$y_3 = 1-(-2)\left(-\frac{3}{5}\right) = -1/5.$$

Calculating x_1, x_2 and x_3

$$x_3 = y_3 / \alpha_3 = -1/5/11/5 = -0.0909,$$

$$x_2 = \left(-2-(2)(-0.0909)\right)/(-5) = 0.3636,$$

$$x_1 = (y_1 - \gamma_1 x_2) / \alpha_1 = \left(1-(2)(0.3636)\right)/(1) = 0.2727.$$

Main steps for solving the banded Doolittle's factorization method (for tridiagonal matrices) in MATLAB:

1) The output will be x and input is n.
2) We define the diagonal, lower diagonal and upper diagonal entries in a loop. Assume a 3 × 3 matrix.

$$\begin{bmatrix} a_1 & c_1 & 0 \\ b_2 & a_2 & c_2 \\ 0 & b_3 & a_3 \end{bmatrix} = \begin{bmatrix} \alpha_1 & \gamma_1 & 0 \\ \beta_2\alpha_1 & \beta_2\gamma_1 + \alpha_2 & \gamma_2 \\ 0 & \beta_3\alpha_2 & \beta_3\gamma_2 + \alpha_3 \end{bmatrix}.$$

   In this case we don't use the zeros command for matrix manipulation.
   As the a's are diagonal entries, the loop starts from 1 and ends at n, b's start from second row so loop starts from 2 and ends at n, c's start from first row and end at n – 1.
3) As α_1 is equal to a_1 so we can write it as
   alpha(1) = a(1).
4) The values of $\beta's$ are equal to $c's$. For this we start a loop from 1 to n – 1 and define $\gamma's$, i.e.,
   gama(j) = c(j).
5) As the first value α_1 is defined so for the remaining we start the loop from 2 to n and implement β values in this loop. As

$$\beta_2 = \frac{b_2}{\alpha_1}, \ \beta_3 = \frac{b_3}{\alpha_2}.$$

   In MATLAB we can write it as
   beta(j) = b(j)/alpha(j – 1), where j = 2 to n.
   And for α's we have the relation
   $\alpha_2 = a_2 - \beta_2\gamma_1, \ \alpha_3 = a_3 - \beta_3\gamma_2.$
   In MATLAB, we can write in journal form as
   alpha(j) = a(j) – beta(j) * gama(j – 1), where j =2 to n.
   Remember both α's and β's are used in one loop.
6) The values of R are defined from 1 to n.
7) The values of y_1, y_2 and y_3 are calculated as the values are
   $y_1 = r_1, \ y_2 = r_2 - y_1\beta_2, \ y_3 = r_3 - y_2\beta_3.$
   In MATLAB define y(1) = r(1).
   And start the loop from 2 to n.
   y(j) = r(j) – y(j – 1) * beta(j).
8) At the end back substitution is performed for the desired values of x.

```
% Mfile
function x=Doolittles_tridiagnol(n)
for j=1:n
  a(j)=1; % Diagonal entries
end
for j=2:n
  b(j)=3; % Lower diagonal entries
end
for j=1:n-1
  c(j)=2; % Upper diagonal entries
end
for j=1:n-1
gama(j)=c(j);
end
alpha(1)=a(1);
for j=2:n
  beta(j)=b(j)/alpha(j-1);
  alpha(j)=a(j)-beta(j)*gama(j-1);
end
for j=1:n
  r(j)=1;
end
y(1)=r(1);
for j=2:n
  y(j)=r(j)-y(j-1)*beta(j);
end
x(n)=y(n)/alpha(n);
for j=n-1:-1:1  % Back substitution UX=Y
  x(j)=(y(j)-gama(j)*x(j+1))/alpha(j);
end
```

```
>> % Command Window
>> Doolittles_tridiagnol(3)
ans =
  0.2727  0.3636  -0.0909
```

2.5.2 CROUT'S FACTORIZATION METHOD FOR BANDED TRIDIAGONAL MATRIX

Consider a 3 × 3 tridiagonal matrix

$$A = \begin{bmatrix} a_1 & c_1 & 0 \\ b_2 & a_2 & c_2 \\ 0 & b_3 & a_3 \end{bmatrix}. \tag{2.56}$$

The matrix A can be written as

$$A = LU.$$

For Crout's method L and U are

$$L = \begin{bmatrix} \alpha_1 & 0 & 0 \\ \beta_2 & \alpha_2 & 0 \\ 0 & \beta_3 & \alpha_3 \end{bmatrix}, U = \begin{bmatrix} 1 & \gamma_1 & 0 \\ 0 & 1 & \gamma_2 \\ 0 & 0 & 1 \end{bmatrix}. \tag{2.57}$$

Multiplication of L and U gives

$$A = \begin{bmatrix} \alpha_1 & \alpha_1\gamma_1 & 0 \\ \beta_2 & \beta_2\gamma_1 + \alpha_2 & \alpha_2\gamma_2 \\ 0 & \beta_3 & \beta_3\gamma_2 + \alpha_3 \end{bmatrix}. \tag{2.58}$$

Comparing both A's

$$\begin{bmatrix} a_1 & c_1 & 0 \\ b_2 & a_2 & c_2 \\ 0 & b_3 & a_3 \end{bmatrix} = \begin{bmatrix} \alpha_1 & \alpha_1\gamma_1 & 0 \\ \beta_2 & \beta_2\gamma_1 + \alpha_2 & \alpha_2\gamma_2 \\ 0 & \beta_3 & \beta_3\gamma_2 + \alpha_3 \end{bmatrix}. \tag{2.59}$$

The diagonal entries are:

$$\alpha_1 = a_1, \alpha_2 = a_2 - \beta_2\gamma_1, \alpha_3 = a_3 - \beta_3\gamma_2. \tag{2.60}$$

The upper diagonal entries are:

$$\gamma_1 = c_1 / \alpha_1, \gamma_2 = c_2 / \alpha_2. \tag{2.61}$$

The lower diagonal entries are:

$$\beta_2 = b_2, \beta_3 = b_3. \tag{2.62}$$

The L and U have the following form

$$L = \begin{bmatrix} \alpha_1 = a_1 & 0 & 0 \\ \beta_2 = b_2 & \alpha_2 = a_2 - \beta_2\gamma_1 & 0 \\ 0 & \beta_3 = b_3 & \alpha_3 = a_3 - \beta_3\gamma_2 \end{bmatrix}, \tag{2.63}$$

$$U = \begin{bmatrix} 1 & \gamma_1 = c_1/\alpha_1 & 0 \\ 0 & 1 & \gamma_2 = c_2/\alpha_2 \\ 0 & 0 & 1 \end{bmatrix}. \tag{2.64}$$

Put

$$LUX = R. \tag{2.65}$$

Calculate $LY = R$

$$\begin{bmatrix} \alpha_1 & 0 & 0 \\ \beta_2 & \alpha_2 & 0 \\ 0 & \beta_3 & \alpha_3 \end{bmatrix}\begin{bmatrix} y_1 \\ y_2 \\ y_3 \end{bmatrix} = \begin{bmatrix} r_1 \\ r_2 \\ r_3 \end{bmatrix}. \tag{2.66}$$

This gives

$$y_1 = r_1 / \alpha_1,$$

$$y_2 = (r_2 - y_1\beta_2)/\alpha_2,$$

$$y_3 = (r_3 - y_2\beta_3)/\alpha_3. \tag{2.67}$$

Finally, put $UX = Y$

$$\begin{bmatrix} 1 & \gamma_1 & 0 \\ 0 & 1 & \gamma_2 \\ 0 & 0 & 1 \end{bmatrix} \begin{bmatrix} x_1 \\ x_2 \\ x_3 \end{bmatrix} = \begin{bmatrix} y_1 \\ y_2 \\ y_3 \end{bmatrix}. \tag{2.68}$$

Solving Eq. (2.68) gives

$$x_3 = y_3,$$

$$x_2 = (y_2 - \gamma_2 x_3),$$

$$x_1 = (y_1 - \gamma_1 x_2). \tag{2.69}$$

Example 2.8: Solve the following tridiagonal system

$$\begin{bmatrix} 1 & 2 & 0 \\ 3 & 1 & 2 \\ 0 & 3 & 1 \end{bmatrix} \begin{bmatrix} x_1 \\ x_2 \\ x_3 \end{bmatrix} = \begin{bmatrix} 1 \\ 1 \\ 1 \end{bmatrix}.$$

Solution:
   Let

$$A = \begin{bmatrix} 1 & 2 & 0 \\ 3 & 1 & 2 \\ 0 & 3 & 1 \end{bmatrix}, R = \begin{bmatrix} 1 \\ 1 \\ 1 \end{bmatrix}.$$

For Crout's tridiagonal matrix

$$\begin{bmatrix} 1 & 2 & 0 \\ 3 & 1 & 2 \\ 0 & 3 & 1 \end{bmatrix} = \begin{bmatrix} \alpha_1 & \alpha_1 \gamma_1 & 0 \\ \beta_2 & \beta_2 \gamma_1 + \alpha_2 & \alpha_2 \gamma_2 \\ 0 & \beta_3 & \beta_3 \gamma_2 + \alpha_3 \end{bmatrix}.$$

Comparing gives

$$\alpha_1 = 1, \gamma_1 = \frac{2}{1} = 2, \beta_2 = 3, \alpha_2 = 1 - (3)(2) = -5,$$

$$\gamma_2 = -\frac{2}{5}, \beta_3 = 3, \alpha_3 = 1 - (3)\left(-\frac{2}{5}\right) = \frac{11}{5} = 2.2.$$

So

$$L = \begin{bmatrix} 1 & 0 & 0 \\ 3 & -5 & 0 \\ 0 & 3 & 11/5 \end{bmatrix}, U = \begin{bmatrix} 1 & 2 & 0 \\ 0 & 1 & -2/5 \\ 0 & 0 & 1 \end{bmatrix}.$$

   Next

$$y_1 = 1,$$

$$y_2 = \left(1 - (1)(3)\right) / -5 = 0.4,$$

$$y_3 = \left(1-(0.4)(3)\right)/2.2 = -0.0909.$$

Calculating x_1, x_2 and x_3 gives

$$x_3 = y_3 = -0.0909,$$

$$x_2 = \left(0.4 - (-0.4)(-0.0909)\right) = 0.3636,$$

$$x_1 = (y_1 - \gamma_1 x_2) = \left(1 - (2)(0.3636)\right) = 0.2727.$$

Main steps for solving banded Crout's factorization method (for tridiagonal matrices) in MATLAB:

1) The output will be x and input is n.
2) Assume a 3 × 3 matrix.

$$\begin{bmatrix} a_1 & c_1 & 0 \\ b_2 & a_2 & c_2 \\ 0 & b_3 & a_3 \end{bmatrix} = \begin{bmatrix} \alpha_1 & \alpha_1\gamma_1 & 0 \\ \beta_2 & \beta_2\gamma_1 + \alpha_2 & \alpha_2\gamma_2 \\ 0 & \beta_3 & \beta_3\gamma_2 + \alpha_3 \end{bmatrix}.$$

The diagonal entries start from 1 and end at n, b's start from second row first entry so the loop starts from 2 and ends at n, c's start from first row and end at n – 1 but as we end at n we ignore n value as α's and γ's both are used in the same loop.
3) As α_1 is equal to a_1 we can write it as
alpha(1) = a(1).
4) The values of $\beta's$ are equal to $b's$. For this we start a loop from 2 to n, i.e.,
beta(j) = b(j).
5) The value of γ_1 is defined for remaining values, we start the loop from 2 to n and both γ and α values are implemented in one loop. As

$$\alpha_2 = a_2 - \beta_2\gamma_1, \alpha_3 = a_3 - \beta_3\gamma_2.$$

And for γ 's we have the relation

$$\gamma_2 = c_2/\alpha_2.$$

In MATLAB, we can write in journal form as
alpha(j) = a(j) – beta(j) * gama(j –1),
gama(j) = c(j)/alpha(j), where j = 2 to n.
6) The values of R are defined from 1 to n.
7) The values of y_1, y_2 and y_3 are calculated, which are
$$y_1 = r_1/\alpha_1, y_2 = (r_2 - y_1\beta_2)/\alpha_2, y_3 = (r_3 - y_2\beta_3)/\alpha_3.$$

In MATLAB, define

y(1) = r(1)/alpha(1).
And start the loop from 2 to n, we have
y(j) = (r(j) – y(j – 1) * beta(j))/alpha(j).
8) At the end back substitution is performed
$$x_3 = y_3, x_2 = (y_2 - \gamma_2 x_3), x_1 = (y_1 - \gamma_1 x_2).$$

In MATLAB we can write it as
x(j) = y(j) – gama(j) * x(j + 1), where j = n – 1 to 1.

```
% Mfile
function x=Crouts_tridiagnol(n)
for j=1:n
  a(j)=1; % Diagonal entries
end
for j=2:n
  b(j)=3; % Lower diagonal entries
end
for j=1:n
  c(j)=2; % Upper diagonal entries
end
alpha(1)=a(1);
for j=2:n
  beta(j)=b(j);
end
gama(1)=c(1)/alpha(1);
for j=2:n
  alpha(j)=a(j)-beta(j)*gama(j-1);
gama(j)=c(j)/alpha(j);
end
for j=1:n
  r(j)=1;
end
y(1)=r(1)/alpha(1);
for j=2:n
 y(j)=(r(j)-y(j-1)*beta(j))/alpha(j);
end
x(n)=y(n);
for j=n-1:-1:1  % Back substitution UX=Y
  x(j)=y(j)-gama(j)*x(j+1);
end
```

```
>> % Command Window
>> x=Crouts_tridiagnol(3)
x =
  0.2727  0.3636  -0.0909
```

2.6   BANDED BLOCK TRIDIAGONAL MATRICES

A tridiagonal matrix which is the collection of small matrices is said to be a block tridiagonal matrix. For example

$$
\begin{bmatrix} A \end{bmatrix} = \begin{bmatrix}
\begin{bmatrix} a_1 \end{bmatrix} & \begin{bmatrix} c_1 \end{bmatrix} & \cdot & & \cdot & & \cdot & & \cdot \\
\begin{bmatrix} b_2 \end{bmatrix} & \begin{bmatrix} a_2 \end{bmatrix} & \begin{bmatrix} c_2 \end{bmatrix} & & \cdot & & & & \cdot \\
\cdot & \cdot & & & \cdot & & \cdot & \\
& \cdot & & & \cdot & & \cdot & \\
\cdot & \cdot & \cdot & \begin{bmatrix} b_{n-1} \end{bmatrix} & \begin{bmatrix} a_{n-1} \end{bmatrix} & \begin{bmatrix} c_{n-1} \end{bmatrix} \\
\cdot & \cdot & \cdot & \cdot & \begin{bmatrix} b_n \end{bmatrix} & \begin{bmatrix} a_n \end{bmatrix}
\end{bmatrix}. \tag{2.70}
$$

The above matrix is an $n \times n$ block tridiagonal matrix with each element of matrix form. The remaining entries of the matrix are zero. The elements can be 2×2, 3×3, etc. matrices. The above tridiagonal system can be decomposed into [A] = [L][U]. In this book, banded Doolittle's and Crout's factorization methods are used to solve such problems.

2.6.1 Doolittle's Factorization Method for Banded Block Tridiagonal Matrix

Consider a 3×3 banded block tridiagonal matrix

$$[A] = \begin{bmatrix} [a_1] & [c_1] & 0 \\ [b_2] & [a_2] & [c_2] \\ 0 & [b_3] & [a_3] \end{bmatrix}. \tag{2.71}$$

The banded block matrix [A] can be written as

$$[A] = [L][U].$$

For Doolittle's method [L] and [U] are

$$[L] = \begin{bmatrix} [I] & 0 & 0 \\ [\beta_2] & [I] & 0 \\ 0 & [\beta_3] & [I] \end{bmatrix}, [U] = \begin{bmatrix} [\alpha_1] & [\gamma_1] & 0 \\ 0 & [\alpha_2] & [\gamma_2] \\ 0 & 0 & [\alpha_3] \end{bmatrix}. \tag{2.72}$$

Where *[I]* is an identity matrix. After implementing the matrix multiplication and comparing both sides we have

The diagonal entries are:

$$[\alpha_1] = [a_1], [\alpha_2] = [a_2] - [\beta_2][\gamma_1], [\alpha_3] = [a_3] - [\beta_3][\gamma_2]. \tag{2.73}$$

The upper diagonal entries are:

$$[\gamma_1] = [c_1], [\gamma_2] = [c_2]. \tag{2.74}$$

The lower diagonal entries are:

$$[\beta_2] = [b_2][\alpha_1]^{-1}, [\beta_3] = [b_3][\alpha_2]^{-1}. \tag{2.75}$$

Note: For block matrices the division is replaced by the inverse of the matrix.

We get [L] and [U] of the following form

$$[L] = \begin{bmatrix} [I] & 0 & 0 \\ [\beta_2] = [b_2][\alpha_1]^{-1} & [I] & 0 \\ 0 & [\beta_3] = [b_3][\alpha_2]^{-1} & [I] \end{bmatrix}, \tag{2.76}$$

$$[U] = \begin{bmatrix} [\alpha_1] = [a_1] & [\gamma_1] = [c_1] & 0 \\ 0 & [\alpha_2] = [a_2] - [\beta_2][\gamma_1] & [\gamma_2] = [c_2] \\ 0 & 0 & [\alpha_3] = [a_3] - [\beta_3][\gamma_2] \end{bmatrix}. \tag{2.77}$$

Put

$$[L][U][X] = [R]. \tag{2.78}$$

Calculate $[L][Y] = [R]$

$$\begin{bmatrix} [I] & 0 & 0 \\ [\beta_2] & [I] & 0 \\ 0 & [\beta_3] & [I] \end{bmatrix} \begin{bmatrix} [y_1] \\ [y_2] \\ [y_3] \end{bmatrix} = \begin{bmatrix} [r_1] \\ [r_2] \\ [r_3] \end{bmatrix}. \tag{2.79}$$

This gives

$$[y_1] = [r_1],$$

$$[y_2] = [r_2] - [\beta_2][y_1],$$

$$[y_3] = [r_3] - [\beta_3][y_2]. \tag{2.80}$$

Finally, $[U][X] = [Y]$

$$\begin{bmatrix} [\alpha_1] & [\gamma_1] & 0 \\ 0 & [\alpha_2] & [\gamma_2] \\ 0 & 0 & [\alpha_3] \end{bmatrix} \begin{bmatrix} [x_1] \\ [x_2] \\ [x_3] \end{bmatrix} = \begin{bmatrix} [y_1] \\ [y_2] \\ [y_3] \end{bmatrix}. \tag{2.81}$$

Solving gives

$$[x_3] = [\alpha_3]^{-1}[y_3],$$

$$[x_2] = [\alpha_2]^{-1}\left([y_2] - [\gamma_2][x_3]\right),$$

$$[x_1] = [\alpha_1]^{-1}\left([y_1] - [\gamma_1][x_2]\right). \tag{2.82}$$

Example 2.9: Solve the following tridiagonal system:

$$A = \begin{bmatrix} \begin{bmatrix} 1 & 0 \\ 0 & 1 \end{bmatrix} & \begin{bmatrix} 0 & 0 \\ 0 & 0 \end{bmatrix} & \begin{bmatrix} 0 & 0 \\ 0 & 0 \end{bmatrix} \\ \begin{bmatrix} 1 & 0 \\ 0 & 1 \end{bmatrix} & \begin{bmatrix} -2 & -0.01 \\ -0.01 & -2 \end{bmatrix} & \begin{bmatrix} 1 & 0 \\ 0 & 1 \end{bmatrix} \\ \begin{bmatrix} 0 & 0 \\ 0 & 0 \end{bmatrix} & \begin{bmatrix} 1 & 0 \\ 0 & 1 \end{bmatrix} & \begin{bmatrix} 1 & 0 \\ 0 & 1 \end{bmatrix} \end{bmatrix}, \quad R = \begin{bmatrix} \begin{bmatrix} 1 \\ 1 \end{bmatrix} \\ \begin{bmatrix} 0 \\ 0 \end{bmatrix} \\ \begin{bmatrix} 0 \\ 0 \end{bmatrix} \end{bmatrix}.$$

Solution:
For Doolittle's tridiagonal matrix

$$\begin{bmatrix} \begin{bmatrix} 1 & 0 \\ 0 & 1 \end{bmatrix} & \begin{bmatrix} 0 & 0 \\ 0 & 0 \end{bmatrix} & \begin{bmatrix} 0 & 0 \\ 0 & 0 \end{bmatrix} \\ \begin{bmatrix} 1 & 0 \\ 0 & 1 \end{bmatrix} & \begin{bmatrix} -2 & -0.01 \\ -0.01 & -2 \end{bmatrix} & \begin{bmatrix} 1 & 0 \\ 0 & 1 \end{bmatrix} \\ \begin{bmatrix} 0 & 0 \\ 0 & 0 \end{bmatrix} & \begin{bmatrix} 1 & 0 \\ 0 & 1 \end{bmatrix} & \begin{bmatrix} 1 & 0 \\ 0 & 1 \end{bmatrix} \end{bmatrix} = \begin{bmatrix} [\alpha_1] & [\gamma_1] & 0 \\ [\beta_2][\alpha_1] & [\beta_2][\gamma_1]+[\alpha_2] & [\gamma_2] \\ 0 & [\beta_3][\alpha_2] & [\beta_3][\gamma_2]+[\alpha_3] \end{bmatrix}.$$

Comparing gives

$$[\gamma_1] = \begin{bmatrix} 0 & 0 \\ 0 & 0 \end{bmatrix}, [\gamma_2] = \begin{bmatrix} 1 & 0 \\ 0 & 1 \end{bmatrix}, [\alpha_1] = \begin{bmatrix} 1 & 0 \\ 0 & 1 \end{bmatrix},$$

$$[\beta_2] = [\alpha_1]^{-1}[b_2] = \begin{bmatrix} 1 & 0 \\ 0 & 1 \end{bmatrix},$$

$$[\alpha_2] = [a_2] - [\beta_2][\gamma_1] = \begin{bmatrix} -2 & -0.01 \\ -0.01 & -2 \end{bmatrix},$$

$$\beta_3 = [b_3][\alpha_2]^{-1} = \begin{bmatrix} -0.5 & 0.0025 \\ 0.0025 & -0.5 \end{bmatrix},$$

$$[\alpha_3] = [a_3] - [\beta_3][\gamma_2] = \begin{bmatrix} 1.5 & -0.0025 \\ -0.0025 & 1.5 \end{bmatrix}.$$

So

$$U = \begin{bmatrix} \begin{bmatrix} 1 & 0 \\ 0 & 1 \end{bmatrix} & \begin{bmatrix} 0 & 0 \\ 0 & 0 \end{bmatrix} & \begin{bmatrix} 0 & 0 \\ 0 & 0 \end{bmatrix} \\ \begin{bmatrix} 0 & 0 \\ 0 & 0 \end{bmatrix} & \begin{bmatrix} -2 & -0.01 \\ -0.01 & -2 \end{bmatrix} & \begin{bmatrix} 1 & 0 \\ 0 & 1 \end{bmatrix} \\ \begin{bmatrix} 0 & 0 \\ 0 & 0 \end{bmatrix} & \begin{bmatrix} 0 & 0 \\ 0 & 0 \end{bmatrix} & \begin{bmatrix} 1.5 & -0.0025 \\ -0.0025 & 1.5 \end{bmatrix} \end{bmatrix}$$

and

$$L = \begin{bmatrix} \begin{bmatrix} 1 & 0 \\ 0 & 1 \end{bmatrix} & \begin{bmatrix} 0 & 0 \\ 0 & 0 \end{bmatrix} & \begin{bmatrix} 0 & 0 \\ 0 & 0 \end{bmatrix} \\ \begin{bmatrix} 1 & 0 \\ 0 & 1 \end{bmatrix} & \begin{bmatrix} 1 & 0 \\ 0 & 1 \end{bmatrix} & \begin{bmatrix} 0 & 0 \\ 0 & 0 \end{bmatrix} \\ \begin{bmatrix} 0 & 0 \\ 0 & 0 \end{bmatrix} & \begin{bmatrix} -0.5 & 0.0025 \\ 0.0025 & -0.5 \end{bmatrix} & \begin{bmatrix} 1 & 0 \\ 0 & 1 \end{bmatrix} \end{bmatrix}.$$

Next

$$[y_1] = [r_1] = \begin{bmatrix} 1 \\ 1 \end{bmatrix},$$

$$[y_2] = \begin{bmatrix} 0 \\ 0 \end{bmatrix} - \begin{bmatrix} 1 & 0 \\ 0 & 1 \end{bmatrix}\begin{bmatrix} 1 \\ 1 \end{bmatrix} = \begin{bmatrix} -1 \\ -1 \end{bmatrix},$$

$$[y_3] = \begin{bmatrix} 0 \\ 0 \end{bmatrix} - \begin{bmatrix} -0.5 & 0.0025 \\ 0.0025 & -0.5 \end{bmatrix}\begin{bmatrix} -1 \\ -1 \end{bmatrix} = \begin{bmatrix} -0.4975 \\ -0.4975 \end{bmatrix}.$$

Calculating x_1, x_2 and x_3 gives

$$[x_3] = [\alpha_3]^{-1}[y_3] = \begin{bmatrix} -0.3324 \\ -0.3322 \end{bmatrix},$$

$$[x_2] = [\alpha_2]^{-1}([y_2] - [\gamma_2][x_3]) = \begin{bmatrix} 0.3321 \\ 0.3322 \end{bmatrix},$$

$$\left[x_1\right] = \left[\alpha_1\right]^{-1}\left(\left[y_1\right] - \left[\gamma_1\right]\left[x_2\right]\right) = \begin{bmatrix} 1 \\ 1 \end{bmatrix}.$$

Main steps for solving banded block Doolittle's factorization method (for tridiagonal matrices) in MATLAB.

1) The domain is defined as:
   xx(i) = xx(i − 1) + h,
   where h and initial value xx(1) are defined above it. The domain is implemented in a loop.
2) Write the diagonal, upper diagonal and lower diagonal entries in the form of a loop as the diagonal entries start from 1 and end at n, upper diagonal entries start from 1 and end at n − 1 and lower diagonal entries start from 2 and end at n. In most cases first and last entries may be different, so we write them separately.
   Note that for block entries we use curly brackets ({}).
3) The diagonal entries excluding the first and final entry are the same. So, we write 2 to n − 1 entries in a loop and the final entry after the loop.
4) As α_1 is equal to a_1 so we can write it as
   alpha{1} = {a1}.
5) The values of $[\gamma]'s$ are equal to $[c]'s$. For this we start a loop from 1 to n − 1 and define $[\gamma]'s$, i.e.,
   gama{j} = c{j}.
6) The first value $[\alpha_1]$ is defined, for remaining values the loop starts from 2 to n along with $[\beta]$ values in the loop. As

$$\left[\beta_2\right] = \frac{\left[b_2\right]}{\left[\alpha_1\right]} \quad , \left[\beta_3\right] = \frac{\left[b_3\right]}{\left[\alpha_2\right]}.$$

In MATLAB we can write
   beta{j} = inv(alpha{j − 1}) * b{j},
and for [α]'s

$$\left[\alpha_2\right] = \left[a_2\right] - \left[\beta_2\right]\left[\gamma_1\right], \left[\alpha_3\right] = \left[a_3\right] - \left[\beta_3\right]\left[\gamma_2\right],$$

we can write in journal form as
   alpha{j} = a{j} − beta{j} * gama{j − 1}, where j = 2 to n.
7) The values of [R] are defined from 1 to n. The values are defined in the form of column vectors.
8) The values of $[y_1]$, $[y_2]$ and $[y_3]$ are calculated as the values are

$$\left[y_1\right] = \left[r_1\right],$$

$$\left[y_2\right] = \left[r_2\right] - \left[y_1\right]\left[\beta_2\right],$$

$$\left[y_3\right] = \left[r_3\right] - \left[y_2\right]\left[\beta_3\right].$$

9) At the end back substitution is performed for the desired values of x.

```
% Mfile
function Block_Band_Doolittles_tridiagnol
n=3;
a{1}=[1 0;0 1];
c{1}=[0 0;0 0];
for j=2:n-1
  a{j}=[-2 -0.01;-0.01 -2];
```

```
end
a{n}=[1 0;0 1];
for j=2:n
  b{j}=[1 0;0 1];
end
for j=2:n-1
  c{j}=[1 0;0 1];
end
r1(1)=1;
  r2(1)=1;
for j=2:n
  r1(j)=0;
  r2(j)=0;
end
for j=1:n-1
gama{j}=c{j};
end
alpha{1}=a{1};
for j=2:n
  beta{j}=inv(alpha{j-1})*b{j};
  alpha{j}=a{j}-beta{j}*gama{j-1};
end
for j=1:n
  rr{j}=[r1(j);r2(j)]
end
y{1}=rr{1};
for j=2:n
  y{j}=rr{j}-beta{j}*y{j-1}
end
x{n}=inv(alpha{n})*y{n};
for j=n-1:-1:1 % Back substitution UX=Y
  x{j}=inv(alpha{j})*(y{j}-gama{j}*x{j+1});
end
```

2.6.2 CROUT'S FACTORIZATION METHOD FOR BANDED BLOCK TRIDIAGONAL MATRIX

Consider a 3 × 3 block banded tridiagonal matrix

$$[A] = \begin{bmatrix} [a_1] & [c_1] & 0 \\ [b_2] & [a_2] & [c_2] \\ 0 & [b_3] & [a_3] \end{bmatrix}. \tag{2.83}$$

The matrix [A] can be written as

$$[A] = [L][U].$$

For Crout's method L and U are

$$[L] = \begin{bmatrix} [\alpha_1] & 0 & 0 \\ [\beta_2] & [\alpha_2] & 0 \\ 0 & [\beta_3] & [\alpha_3] \end{bmatrix}, [U] = \begin{bmatrix} [I] & [\gamma_1] & 0 \\ 0 & [I] & [\gamma_2] \\ 0 & 0 & [I] \end{bmatrix}. \tag{2.84}$$

Here [I] is the identity matrix. Multiply [L] and [U] and compare both A's.

The diagonal entries are:

$$[\alpha_1] = [a_1], [\alpha_2] = [a_2] - [\beta_2][\gamma_1], [\alpha_3] = [a_3] - [\beta_3][\gamma_2]. \tag{2.85}$$

The upper diagonal entries are:

$$[\gamma_1] = [c_1][\alpha_1]^{-1}, [\gamma_2] = [c_2][\alpha_2]^{-1}. \tag{2.86}$$

The lower diagonal entries are:

$$[\beta_2] = [b_2], \ [\beta_3] = [b_3]. \tag{2.87}$$

The [L] and [U] have the following form:

$$[L] = \begin{bmatrix} [\alpha_1] = [a_1] & 0 & 0 \\ [\beta_2] = [b_2] & [\alpha_2] = [a_2] - [\beta_2][\gamma_1] & 0 \\ 0 & [\beta_3] = [b_3] & [\alpha_3] = [a_3] - [\beta_3][\gamma_2] \end{bmatrix}, \tag{2.88}$$

$$U = \begin{bmatrix} [I] & [\gamma_1] = [c_1][\alpha_1]^{-1} & 0 \\ 0 & [I] & [\gamma_2] = [c_2][\alpha_2]^{-1} \\ 0 & 0 & [I] \end{bmatrix}. \tag{2.89}$$

Put

$$[L][U][X] = [R]. \tag{2.90}$$

Calculate $[L][Y] = [R]$

$$\begin{bmatrix} [\alpha_1] & 0 & 0 \\ [\beta_2] & [\alpha_2] & 0 \\ 0 & [\beta_3] & [\alpha_3] \end{bmatrix} \begin{bmatrix} [y_1] \\ [y_2] \\ [y_3] \end{bmatrix} = \begin{bmatrix} [r_1] \\ [r_2] \\ [r_3] \end{bmatrix}. \tag{2.91}$$

This gives

$$[y_1] = [\alpha_1]^{-1}[r_1],$$

$$[y_2] = [\alpha_2]^{-1}([r_2] - [\beta_2][y_1]),$$

$$[y_3] = [\alpha_3]^{-1}([r_3] - [\beta_3][y_2]). \tag{2.92}$$

Finally, $[U][X] = [Y]$

$$\begin{bmatrix} [I] & [\gamma_1] & 0 \\ 0 & [I] & [\gamma_2] \\ 0 & 0 & [I] \end{bmatrix} \begin{bmatrix} [x_1] \\ [x_2] \\ [x_3] \end{bmatrix} = \begin{bmatrix} [y_1] \\ [y_2] \\ [y_3] \end{bmatrix}. \tag{2.93}$$

Solving gives

$$[x_3] = [y_3],$$

$$[x_2] = ([y_2] - [\gamma_2][x_3]),$$

$$[x_1] = ([y_1] - [\gamma_1][x_2]). \tag{2.94}$$

Example 2.10: Solve the following tridiagonal system:

$$A = \begin{bmatrix} \begin{bmatrix} 1 & 0 \\ 0 & 1 \end{bmatrix} & \begin{bmatrix} 0 & 0 \\ 0 & 0 \end{bmatrix} & \begin{bmatrix} 0 & 0 \\ 0 & 0 \end{bmatrix} \\ \begin{bmatrix} 1 & 0 \\ 0 & 1 \end{bmatrix} & \begin{bmatrix} -2 & -0.01 \\ -0.01 & -2 \end{bmatrix} & \begin{bmatrix} 1 & 0 \\ 0 & 1 \end{bmatrix} \\ \begin{bmatrix} 0 & 0 \\ 0 & 0 \end{bmatrix} & \begin{bmatrix} 1 & 0 \\ 0 & 1 \end{bmatrix} & \begin{bmatrix} 1 & 0 \\ 0 & 1 \end{bmatrix} \end{bmatrix}, R = \begin{bmatrix} 1 \\ 1 \\ 0 \\ 0 \\ 0 \\ 0 \end{bmatrix}.$$

Solution:

For Crout's tridiagonal matrix

$$\begin{bmatrix} \begin{bmatrix} 1 & 0 \\ 0 & 1 \end{bmatrix} & \begin{bmatrix} 0 & 0 \\ 0 & 0 \end{bmatrix} & \begin{bmatrix} 0 & 0 \\ 0 & 0 \end{bmatrix} \\ \begin{bmatrix} 1 & 0 \\ 0 & 1 \end{bmatrix} & \begin{bmatrix} -2 & -0.01 \\ -0.01 & -2 \end{bmatrix} & \begin{bmatrix} 1 & 0 \\ 0 & 1 \end{bmatrix} \\ \begin{bmatrix} 0 & 0 \\ 0 & 0 \end{bmatrix} & \begin{bmatrix} 1 & 0 \\ 0 & 1 \end{bmatrix} & \begin{bmatrix} 1 & 0 \\ 0 & 1 \end{bmatrix} \end{bmatrix} = \begin{bmatrix} [\alpha_1] & [\alpha_1][\gamma_1] & 0 \\ [\beta_2] & [\beta_2][\gamma_1] + [\alpha_2] & [\alpha_2][\gamma_2] \\ 0 & [\beta_3] & [\beta_3][\gamma_2] + [\alpha_3] \end{bmatrix}.$$

Comparing both gives

$$[\alpha_1] = \begin{bmatrix} 1 & 0 \\ 0 & 1 \end{bmatrix}, [\beta_2] = [b_2] = \begin{bmatrix} 1 & 0 \\ 0 & 1 \end{bmatrix}, \beta_3 = [b_3] = \begin{bmatrix} 1 & 0 \\ 0 & 1 \end{bmatrix}, [\gamma_1] = [\alpha_1]^{-1}[c_1] = \begin{bmatrix} 0 & 0 \\ 0 & 0 \end{bmatrix},$$

$$[\alpha_2] = [a_2] - [\beta_2][\gamma_1] = \begin{bmatrix} -2 & -0.01 \\ -0.01 & -2 \end{bmatrix}, [\gamma_2] = [\alpha_2]^{-1}[c_2] = \begin{bmatrix} -0.5 & 0.0025 \\ 0.0025 & -0.5 \end{bmatrix},$$

$$[\alpha_3] = [a_3] - [\beta_3][\gamma_2] = \begin{bmatrix} 1.5 & -0.0025 \\ -0.0025 & 1.5 \end{bmatrix},$$

so

$$L = \begin{bmatrix} \begin{bmatrix} 1 & 0 \\ 0 & 1 \end{bmatrix} & \begin{bmatrix} 0 & 0 \\ 0 & 0 \end{bmatrix} & \begin{bmatrix} 0 & 0 \\ 0 & 0 \end{bmatrix} \\ \begin{bmatrix} 1 & 0 \\ 0 & 1 \end{bmatrix} & \begin{bmatrix} -2 & -0.01 \\ -0.01 & -2 \end{bmatrix} & \begin{bmatrix} 0 & 0 \\ 0 & 0 \end{bmatrix} \\ \begin{bmatrix} 0 & 0 \\ 0 & 0 \end{bmatrix} & \begin{bmatrix} 1 & 0 \\ 0 & 1 \end{bmatrix} & \begin{bmatrix} 1.5 & -0.0025 \\ -0.0025 & 1.5 \end{bmatrix} \end{bmatrix}$$

and

$$U = \begin{bmatrix} \begin{bmatrix} 1 & 0 \\ 0 & 1 \end{bmatrix} & \begin{bmatrix} 0 & 0 \\ 0 & 0 \end{bmatrix} & \begin{bmatrix} 0 & 0 \\ 0 & 0 \end{bmatrix} \\ \begin{bmatrix} 0 & 0 \\ 0 & 0 \end{bmatrix} & \begin{bmatrix} 1 & 0 \\ 0 & 1 \end{bmatrix} & \begin{bmatrix} -0.5 & 0.0025 \\ 0.0025 & -0.5 \end{bmatrix} \\ \begin{bmatrix} 0 & 0 \\ 0 & 0 \end{bmatrix} & \begin{bmatrix} 0 & 0 \\ 0 & 0 \end{bmatrix} & \begin{bmatrix} 1 & 0 \\ 0 & 1 \end{bmatrix} \end{bmatrix}.$$

Next

$$[y_1] = [\alpha_1]^{-1}[r_1] = \begin{bmatrix} 1 \\ 1 \end{bmatrix},$$

$$[y_2] = [\alpha_2]^{-1}([r_2] - [\beta_2][y_1]) = \begin{bmatrix} 0.4975 \\ 0.4975 \end{bmatrix},$$

$$[y_3] = [\alpha_3]^{-1}([r_3] - [\beta_3][y_2]) = \begin{bmatrix} -0.3322 \\ -0.3322 \end{bmatrix}.$$

Calculating x_1, x_2 and x_3

$$[x_3] = [y_3] = \begin{bmatrix} -0.3322 \\ -0.3322 \end{bmatrix},$$

$$[x_2] = ([y_2] - [\gamma_2][x_3]) = \begin{bmatrix} 0.3322 \\ 0.3322 \end{bmatrix},$$

$$[x_1] = ([y_1] - [\gamma_1][x_2]) = \begin{bmatrix} 1 \\ 1 \end{bmatrix}.$$

Main steps for solving banded block Crout's factorization method (for tridiagonal matrices) in MATLAB:

1) The block elements are adjusted in the same style.
2) As $[\alpha_1]$ is equal to $[a_1]$ so we can write it as
   alpha{1} = a{1}.
3) The values of $\beta's$ are equal to $b's$. For this we start a loop from 2 to n.
   beta{j} = b{j}.
4) The value of $[\gamma_1]$ is defined for remaining values, we start the loop from 2 to n for both $[\gamma]$ and $[\alpha]$ in the loop. As

$$[\alpha_2] = [a_2] - [\beta_2][\gamma_1], [\alpha_3] = [a_3] - [\beta_3][\gamma_2],$$

and for $[\gamma]$'s

$$[\gamma_2] = [\alpha_2]^{-1}[c_2].$$

In MATLAB, we can write in journal form as
   a{j} = a{j} − (b{j} * gamma{j − 1}),
   gamma{j} = inv(a{j}) * c{j}, where j = 2 to n.
5) Define R from 1 to n.
6) The values of $[y_1]$, $[y_2]$ and $[y_3]$ are calculated.
7) At the end back substitution is performed.

```
% Mfile
function x=Crouts_tridiagnol_banded
n=3;
a{1}=[1 0;0 1];
c{1}=[0 0;0 0];
for j=2:n-1
  a{j}=[-2 -0.01;-0.01 -2];
end
a{n}=[1 0;0 1];
```

```
for j=2:n
  b{j}=[1 0;0 1];
end
for j=2:n
  c{j}=[1 0;0 1];
end
alpha{1}=a{1};
for j=2:n
  beta{j}=b{j};
end
gama{1}=c{1}*inv(alpha{1});
for j=2:n
  alpha{j}=a{j}-beta{j}*gama{j-1};
gama{j}=c{j}*inv(alpha{j});
end
r1(1)=1;
  r2(1)=1;
for j=2:n
  r1(j)=0;
  r2(j)=0;
end
for j=1:n
  rr{j}=[r1(j);r2(j)];
end
y{1}=inv(a{1})*rr{1};
for j=2:n
 y{j}=inv(alpha{j})*(rr{j}-beta{j}*y{j-1});
end
x{n}=y{n};
for j=n-1:-1:1 % Back substitution UX=Y
  x{j}=y{j}-gama{j}*x{j+1};
end
```

2.7 ITERATIVE METHODS

Previously, the methods used to solve the system of linear algebraic equations were direct. Now indirect/iterative methods are used to solve such systems. The iterative methods are not always successful for systems of linear algebraic equations. These methods will be successful if and only if the diagonal elements of the coefficient matrix are larger than the sum of other elements. Consider a 3×3 system:

$$a_{11}x_1 + a_{12}x_2 + a_{13}x_3 = b_1,$$

$$a_{21}x_1 + a_{22}x_2 + a_{23}x_3 = b_2,$$

$$a_{31}x_1 + a_{32}x_2 + a_{33}x_3 = b_3. \tag{2.95}$$

Eq. (2.95) can be solved by an iterative method if

$$|a_{11}| > |a_{12}| + |a_{13}|,$$

$$|a_{22}| > |a_{21}| + |a_{23}|,$$

$$|a_{33}| > |a_{31}| + |a_{32}|. \tag{2.96}$$

2.7.1 GAUSS-JACOBI METHOD

For this method consider a system of three linear equations

$$a_{11}x_1 + a_{12}x_2 + a_{13}x_3 = b_1,$$

$$a_{21}x_1 + a_{22}x_2 + a_{23}x_3 = b_2,$$

$$a_{31}x_1 + a_{32}x_2 + a_{33}x_3 = b_3. \tag{2.97}$$

With

$$|a_{11}| > |a_{12}| + |a_{13}|,$$

$$|a_{22}| > |a_{21}| + |a_{23}|,$$

$$|a_{33}| > |a_{31}| + |a_{32}|. \tag{2.98}$$

For the Gauss-Jacobi method, eliminate the diagonal elements from each equation, i.e.,

$$x^{j+1}_1 = (b_1 - (a_{12}x^j_2 + a_{13}x^j_3))/a_{11},$$

$$x^{j+1}_2 = (b_2 - (a_{21}x^j_1 + a_{23}x^j_3))/a_{22},$$

$$x^{j+1}_3 = (b_3 - (a_{31}x^j_1 + a_{32}x^j_2))/a_{33}. \tag{2.99}$$

If x^0_1, x^0_2 and x^0_3 are the initial values of x_1, x_2 and x_3, then for $j = 0$ we have

$$x^1_1 = (b_1 - (a_{12}x^0_2 + a_{13}x^0_3))/a_{11},$$

$$x^1_2 = (b_2 - (a_{21}x^0_1 + a_{23}x^0_3))/a_{22},$$

$$x^1_3 = (b_3 - (a_{31}x^0_1 + a_{32}x^0_2))/a_{33}. \tag{2.100}$$

Using the values x^1_1, x^1_2 and x^1_3 for the next iteration (for $j = 1$) we have

$$x^2_1 = (b_1 - (a_{12}x^1_2 + a_{13}x^1_3))/a_{11},$$

$$x^2_2 = (b_2 - (a_{21}x^1_1 + a_{23}x^1_3))/a_{22},$$

$$x^2_3 = (b_3 - (a_{31}x^1_1 + a_{32}x^1_2))/a_{33}. \tag{2.101}$$

The values $x^2{}_1$, $x^2{}_2$ and $x^2{}_3$ are used for the next iteration (for j = 2) and the procedure is stopped when we get repeated values.

The second form is much better than the first one. It is guaranteed if the system is diagonally dominant then there is no need to calculate the convergence. However, the convergence is often calculated for weaker diagonally dominant systems.

Example 2.11: Solve the system by using the Gauss-Jacobi method:

$$10x_1 - 5x_2 - 2x_3 = 3,$$
$$4x_1 - 10x_2 + 3x_3 = -3,$$
$$x_1 + 6x_2 + 10x_3 = -3.$$

Solution: As can be seen this system is diagonally dominant. For the Gauss-Jacobi method, write the system as:

$$x^{j+1}{}_1 = \left(3 + 5x^j{}_2 + 2x^j{}_3\right)/10,$$

$$x^{j+1}{}_2 = \left(3 + 4x^j{}_1 + 3x^j{}_3\right)/10,$$

$$x^{j+1}{}_3 = \left(-3 - x^j{}_1 - 6x^j{}_2\right)/10.$$

Where j = 0,1,2,3 ...

Assume that initial values are (0, 0, 0). For j = 0, the required output will be

$$x^1{}_1 = \left(3 + 5(0) + 2(0)\right)/10 = 0.3,$$

$$x^1{}_2 = \left(3 + 4(0) + 3(0)\right)/10 = 0.3,$$

$$x^1{}_3 = \left(-3 - (0) - 6(0)\right)/10 = -0.3.$$

Using the second iteration (j = 1)

$$x^2{}_1 = \left(3 + 5(0.3) + 2(-0.3)\right)/10 = 0.39,$$

$$x^2{}_2 = \left(3 + 4(0.3) + 3(-0.3)\right)/10 = 0.33,$$

$$x^2{}_3 = \left(-3 - (0.3) - 6(0.3)\right)/10 = -0.53.$$

Using the third iteration (j = 2)

$$x^3{}_1 = \left(3 + 5(0.33) + 2(-0.53)\right)/10 = 0.36,$$

$$x^3{}_2 = \left(3 + 4(0.39) + 3(-0.53)\right)/10 = 0.30,$$

$$x^3{}_3 = \left(-3 - (0.39) - 6(0.33)\right)/10 = -0.53.$$

Using the fourth iteration (j = 3)

$$x^4{}_1 = \left(3 + 5(0.30) + 2(-0.53)\right)/10 = 0.34,$$

$$x^4{}_2 = \left(3 + 4(0.36) + 3(-0.53)\right)/10 = 0.28,$$

$$x^4{}_3 = \left(-3 - (0.36) - 6(0.30)\right)/10 = -0.52.$$

By repeating the same procedure till $j = 6$ we will get the correct values up to two decimal places which are:

$$x_1 = 0.34,\ x_2 = 0.28,\ x_3 = -0.50.$$

Main steps for solving the Gauss-Jacobi method in MATLAB:

1) The inputs and outputs are defined first. The solution X and number of iterations will be outputs, the inputs are matrix, vector b and iterations used in each loop.
2) Length of the matrix and b will be the same, i.e., n = length (b).
3) b must be a column vector, i.e., b = b'.
4) L must be a column vector, i.e., L = L'.
5) The solution X has n rows and one column, i.e., X = zeros(n,1).
6) Two loops are started, one for upper iterations j's and other for lower iterations i's.

   Note that in the above example, the first equation on the right side $a_{11}x^j{}_1$ is excluded. Similarly, in the second equation $a_{22}x^j{}_2$ is excluded and in the same manner in the third one $a_{33}x^j{}_3$ is excluded, i.e.,

$$x^{j+1}{}_1 = \left(3 + 5x^j{}_2 + 2x^j{}_3\right)/10, \qquad x^j{}_1 ??$$

$$x^{j+1}{}_2 = \left(3 + 4x^j{}_1 + 3x^j{}_3\right)/10, \quad x^j{}_2 ??$$

$$x^{j+1}{}_3 = \left(-3 - x^j{}_1 - 6x^j{}_2\right)/10, \quad x^j{}_3 ??$$

In MATLAB, it can be defined as

X(i) = (b(i) − A(i,[1:i − 1,i + 1:n]) * L([1:i − 1,i + 1:n]))/A(i,i), where i = 1 to n.

7) Next X is replaced with L because the previous iteration is replaced with the current one.

```
% Mfile
function [X,numiter]=Gauss_Jacobi(A,b,L)   % Outputs X and numiter
%and inputs are A, L and b
n=length(b); % Length must be same
b=b';  % Column vector
L=L';   % Column vector
X=zeros(n,1); % X have n rows and one column
for numiter=1:100
  for i=1:n
   X(i)=(b(i)-A(i,[1:i-1,i+1:n])*L([1:i-1,i+1:n]))/A(i,i);
   L=X;
  end
end
```

```
>> % Command Window
>> [X]=Gauss_Jacobi([10 -5 -2;4 -10 3;1 6 10],[3 -3 -3],[0 0 0])
X =
  0.3415
  0.2850
 -0.5052
```

2.7.2 GAUSS-SEIDEL METHOD

In the Gauss-Seidel method, Eq. (2.97) can be written as

$$x^{j+1}_1 = \left(b_1 - \left(a_{12}x^j_2 + a_{13}x^j_2\right)\right)\big/a_{11},$$

$$x^{j+1}_2 = \left(b_2 - \left(a_{21}x^{j+1}_1 + a_{23}x^j_3\right)\right)/a_{22},$$

$$x^{j+1}_3 = \left((b_3 - \left(a_{31}x^{j+1}_1 + a_{32}x^{j+1}_2\right)\right)/a_{33}. \tag{2.102}$$

If x^0_1, x^0_2 and x^0_3 are the initial values of x_1, x_2 and x_3, then

$$x^1_1 = \left(b_1 - \left(a_{12}x^0_2 + a_{13}x^0_3\right)\right)/a_{11},$$

$$x^1_2 = \left(b_2 - \left(a_{21}x^1_1 + a_{23}x^0_3\right)\right)/a_{22},$$

$$x^1_3 = \left((b_3 - \left(a_{31}x^1_1 + a_{32}x^1_2\right)\right)/a_{33}. \tag{2.103}$$

The procedure is stopped when we get repeated values.

Example 2.12: Solve the system by using the Gauss-Seidel method:

$$10x_1 - 5x_2 - 2x_3 = 3,$$

$$4x_1 - 10x_2 + 3x_3 = -3,$$

$$x_1 + 6x_2 + 10x_3 = -3.$$

Solution: As can be seen this system is diagonally dominant. For the Gauss-Seidel method, write the system as:

$$x^{j+1}_1 = \left(3 + 5x^j_2 + 2x^j_3\right)/10,$$

$$x^{j+1}_2 = \left(3 + 4x^{j+1}_1 + 3x^j_3\right)/10,$$

$$x^{j+1}_3 = \left(-3 - x^{j+1}_1 - 6x^{j+1}_2\right)/10.$$

Where j = 0,1,2,3 …
Assume that initial values are (0,0,0). For j = 0, the required output will be

$$x^1_1 = \left(3 + 5(0) + 2(0)\right)/10 = 0.3,$$

$$x^1_2 = \left(3 + 4(0.3) + 3(0)\right)/10 = 0.42,$$

$$x^1_3 = \left(-3 - (0.3) - 6(0.42)\right)/10 = -0.58.$$

Using the second iteration (j = 1)

$$x^2_1 = \left(3 + 5(0.42) + 2(-0.58)\right)/10 = 0.39,$$

$$x^2{}_2 = \left(3 + 4(0.39) + 3(-0.58)\right)/10 = 0.28,$$

$$x^2{}_3 = \left(-3 - (0.39) - 6(0.28)\right)/10 = -0.50.$$

Using the third iteration (j = 2)

$$x^3{}_1 = \left(3 + 5(0.28) + 2(-0.50)\right)/10 = 0.33,$$

$$x^3{}_2 = \left(3 + 4(0.33) + 3(-0.50)\right)/10 = 0.28,$$

$$x^3{}_3 = \left(-3 - (0.33) - 6(0.28)\right)/10 = -0.50.$$

The process is stopped for j = 6 as the values are repeated for j = 6. The correct values are x_1 = 0.34, x_2 = 0.28 and x_3 = −0.50.

Main steps for solving the Gauss-Seidel method in MATLAB:

1) Define inputs and outputs. The solution X and the number of iterations will be output and matrix A, vector b and iteration L will be input.
2) Length of matrix and b will be the same, i.e., n = length (b).
3) b is a column vector, i.e., b = b'.
4) L is a column vector, i.e., L = L'.
5) The solution X has n rows and one column, i.e., X = zeros(n,1).
6) Two loops are started, one for upper iterations j's and the other for lower iterations i's.
7) Note that in the above example, the first equation on the right side $a_{11}x^k{}_1$ is missing. Similarly, in the second equation $a_{22}x^k{}_2$ is missing and in the same manner in the third one $a_{33}x^k{}_3$ is missing, i.e.,

$$x^{k+1}{}_1 = \left(3 + 5x^k{}_2 + 2x^k{}_3\right)/10, \quad x^k{}_1 ??$$

$$x^{k+1}{}_2 = \left(3 + 4x^{k+1}{}_1 + 3x^k{}_3\right)/10, \quad x^k{}_2 ??$$

$$x^{k+1}{}_3 = \left(-3 - x^{k+1}{}_1 - 6x^{k+1}{}_2\right)/10, \quad x^k{}_3 ??$$

In MATLAB, we exclude the ith term, i.e.,

X(i) = (b(i) − A(i,[1:i − 1,i + 1:n]) * L([1:i − 1,i + 1:n]))/A(i,i), where i = 1 to n.

8) Moreover, X(i) is replaced with L(i) because the previous equation result is added in the next equation and vice versa.

```
% Mfile
function [X,numiter]=Gauss_Seidel(A,b,L)
n=length(b); % Length must be same
b=b'; % Column vector
L=L';
X=zeros(n,1); % X have n rows and one column
for numiter=1:100
  for i=1:n
    X(i)=(b(i)-A(i,[1:i-1,i+1:n])*L([1:i-1,i+1:n]))/A(i,i);
    L(i)=X(i);
  end
end
```

```
>> % Command Window
>> [X]=Gauss_Seidel([10 -5 -2;4 -10 3;1 6 10],[3 -3 -3],[0 0 0])
X =
    0.3415
    0.2850
   -0.5052
```

2.7.3 Conjugate Gradient Method

In this method vector x is calculated which minimizes the scalar function

$$F(x) = 0.5x^t Ax - b^t x. \tag{2.104}$$

The matrix A is symmetric. The gradient $\nabla F = Ax - b$ is equal to zero when function is minimized.

$$Ax = b. \tag{2.105}$$

Minimization in iteration is achieved by the gradient method. Starting with the initial guess x_0, each cycle computes a better solution.

$$x_{k+1} = x_k + (alpha)_k s_k. \tag{2.106}$$

$F(x_{k+1})$ is minimized by x_{k+1} by choosing step length $(alpha)_k$.
   That is, it satisfies (2.105)

$$A\left(x_k + (alpha)_k s_k\right) = b. \tag{2.107}$$

Define the residual

$$r_k = b - Ax_k. \tag{2.108}$$

Eq. (2.107) becomes $(alpha)_k As_k = r_k$. Multiplying both sides by s_k^t and calculating $(alpha)_k$ gives

$$(alpha)_k = \frac{s_k^t r_k}{s_k^t As_k}. \tag{2.109}$$

Now search direction s_k is calculated by the method of steepest descent. Initially $s_k = r_k$ is chosen, the search direction has form

$$s_{k+1} = r_k + (beta)_k s_k. \tag{2.110}$$

The constant $(beta)_k$ is chosen in such a way that two successive directions are noninterfering to each other, i.e., $s_{k+1}^t As_k = 0$. Putting Eq. (2.110) in this yields

$$\left(r_k + (beta)_k s_k\right) As_k = 0. \tag{2.111}$$

Simplifying gives

$$\left(beta\right)_k = -\frac{r_{k+1}^t As_k}{s_k^t As_k}.$$ (2.112)

Example 2.13: Solve the system by using the Conjugate gradient method

$$10x_1 - 5x_2 - 2x_3 = 1,$$

$$4x_1 - 10x_2 + 3x_3 = -1,$$

$$x_1 + 6x_2 + 10x_3 = -1.$$

Solution: Start with the initial guess $x = [0\ 0\ 0]^t$. For $k = 0$

$$r_0 = b - Ax_0 = \begin{bmatrix} 1 \\ -1 \\ -1 \end{bmatrix} - \begin{bmatrix} 10 & -5 & -2 \\ 4 & -10 & 3 \\ 1 & 6 & 10 \end{bmatrix} \begin{bmatrix} 0 \\ 0 \\ 0 \end{bmatrix},$$

$$s_0 = r_0 = \begin{bmatrix} 1 \\ -1 \\ -1 \end{bmatrix},$$

$$As_0 = \begin{bmatrix} 17 \\ 11 \\ -15 \end{bmatrix},$$

So

$$\left(alpha\right)_0 = \frac{s_0^t r_0}{s_0^t As_0} = 0.1429.$$

Putting these values in Eq. (2.106)

$$x_1 = x_0 + \left(alpha\right)_0 s_0 = \begin{bmatrix} 0 \\ 0 \\ 0 \end{bmatrix} + 0.1429 \begin{bmatrix} 1 \\ -1 \\ -1 \end{bmatrix} = \begin{bmatrix} 0.1429 \\ -0.1429 \\ -0.1429 \end{bmatrix}.$$

The value of x_1 is substituted in Eq. (2.108)

$$r_1 = b - Ax_1 = \begin{bmatrix} 1 \\ -1 \\ -1 \end{bmatrix} - \begin{bmatrix} 10 & -5 & -2 \\ 4 & -10 & 3 \\ 1 & 6 & 10 \end{bmatrix} \begin{bmatrix} 0.1429 \\ -0.1429 \\ -0.1429 \end{bmatrix} = \begin{bmatrix} -1.4286 \\ -2.5714 \\ 1.1429 \end{bmatrix}.$$

Eq. (2.112) becomes

$$\left(beta\right)_0 = -\frac{r_1^t As_0}{s_0^t As_0} = 3.3197.$$

Eq. (2.110) becomes

$$s_1 = r_1 + \left(beta\right)_0 s_0 = \begin{bmatrix} -1.4286 \\ -2.5714 \\ 1.1429 \end{bmatrix} + 3.3197 \begin{bmatrix} 1 \\ -1 \\ -1 \end{bmatrix} = \begin{bmatrix} 1.8912 \\ -5.8912 \\ -2.1769 \end{bmatrix}.$$

As

$$As_1 = \begin{bmatrix} 52.7211 \\ 59.9456 \\ -55.2245 \end{bmatrix}.$$

For k = 1:
  Eq. (2.109) becomes

$$(alpha)_1 = \frac{s_1^t r_1}{s_1^t As_1} = -0.0340.$$

Eq. (2.108) becomes

$$x_2 = x_1 + (alpha)_1 \, s_1 = \begin{bmatrix} 0.1429 \\ -0.1429 \\ -0.1429 \end{bmatrix} - 0.0340 \begin{bmatrix} 1.8912 \\ -5.8912 \\ -2.1769 \end{bmatrix} = \begin{bmatrix} 0.0785 \\ 0.0577 \\ -0.0688 \end{bmatrix}.$$

Eq. (2.158) becomes

$$r_2 = b - Ax_2 = \begin{bmatrix} 1 \\ -1 \\ -1 \end{bmatrix} - \begin{bmatrix} 10 & -5 & -2 \\ 4 & -10 & 3 \\ 1 & 6 & 10 \end{bmatrix} \begin{bmatrix} 0.0785 \\ 0.0577 \\ -0.0688 \end{bmatrix} = \begin{bmatrix} 0.3660 \\ -0.5309 \\ -0.7370 \end{bmatrix}.$$

Eq. (2.112) becomes

$$(beta)_1 = -\frac{r_2^t As_1}{s_1^t As_1} = 0.2115.$$

Eq. (2.110) becomes

$$s_2 = r_2 + (beta)_1 \, s_1 = \begin{bmatrix} 0.3660 \\ -0.5309 \\ -0.7370 \end{bmatrix} + 0.2115 \begin{bmatrix} 1.8912 \\ -5.8912 \\ -2.1769 \end{bmatrix} = \begin{bmatrix} 0.7659 \\ -1.7766 \\ -1.1973 \end{bmatrix}.$$

As

$$As_2 = \begin{bmatrix} 18.9370 \\ 17.2380 \\ -21.8665 \end{bmatrix}.$$

For k = 2:
  Eq. (2.109) becomes

$$(alpha)_2 = \frac{s_2^t r_2}{s_2^t As_2} = 0.0690.$$

Eq. (2.106) becomes

$$x_3 = x_2 + (alpha)_2 \, s_2 = \begin{bmatrix} 0.0785 \\ 0.0577 \\ -0.0688 \end{bmatrix} + 0.0690 \begin{bmatrix} 0.7659 \\ -1.7766 \\ -1.1973 \end{bmatrix} = \begin{bmatrix} 0.1313 \\ -0.0649 \\ -0.1514 \end{bmatrix}.$$

Eq. (2.108) becomes

$$r_3 = b - Ax_3 = \begin{bmatrix} 1 \\ -1 \\ -1 \end{bmatrix} - \begin{bmatrix} 10 & -5 & -2 \\ 4 & -10 & 3 \\ 1 & 6 & 10 \end{bmatrix} \begin{bmatrix} 0.1313 \\ -0.0649 \\ -0.1514 \end{bmatrix} = \begin{bmatrix} -0.9409 \\ -1.7206 \\ 0.7721 \end{bmatrix}.$$

Eq. (2.112) becomes

$$\left(beta\right)_2 = -\frac{r_3^t A s_2}{s_2^t A s_2} = 6.3980.$$

Eq. (2.110) becomes

$$s_3 = r_3 + \left(beta\right)_2 s_2 = \begin{bmatrix} -0.9409 \\ -1.7206 \\ 0.7721 \end{bmatrix} + 6.3980 \begin{bmatrix} 0.7659 \\ -1.7766 \\ -1.1973 \end{bmatrix} = \begin{bmatrix} 3.9596 \\ -13.0873 \\ -6.8881 \end{bmatrix}.$$

Eq. (2.109) becomes

$$\left(alpha\right)_3 = \frac{s_3^t r_3}{s_3^t A s_3} = 0.0690.$$

Eq. (2.106) becomes

$$x_4 = x_3 + \left(alpha\right)_3 s_3 = \begin{bmatrix} 0.1313 \\ -0.0649 \\ -0.1514 \end{bmatrix} + 0.0690 \begin{bmatrix} 3.9596 \\ -13.0873 \\ -6.8881 \end{bmatrix} = \begin{bmatrix} 0.1313 \\ -0.0649 \\ -0.1514 \end{bmatrix}.$$

So the correct value is $[0.1313\ -0.0649\ -0.1514]^t$.

Main steps for solving the conjugate gradient method in MATLAB:

1) The inputs and outputs are defined first. The solution x and the number of iterations will be output and matrix A, vector b and initial guess x will be input.
2) Length of matrix and b will be the same, i.e., n = length (b).
3) b must be a column vector, i.e., b = b'.
4) r will be a column vector.
5) For the first iteration s = r.
6) For row vector transpose is taken for this matrix.
7) Next loop will be started for desired output in which iterative formulas will be implemented.
8) First $\left(alpha\right)_k$ will be defined which is

$$\left(alpha\right)_k = \frac{s_k^t r_k}{s_k^t A s_k}.$$

In MATLAB s_k^t is represented by p and $A s_k$ is represented by t.
9) The value of $\left(alpha\right)_k$ is used in

$$x_{k+1} = x_k + \left(alpha\right)_k s_k.$$

10) Again, r is defined, because the first value is already used, for next values the formula of r is written again.
11) Next $\left(beta\right)_k$ is defined:

$$\left(beta\right)_k = -\frac{r_{k+1}^t A s_k}{s_k^t A s_k}.$$

12) This value is used in

$$s_{k+1} = r_k + \left(beta\right)_k s_k.$$

13) Next the loop will be ended, and the program will be run on the command window.

```
% Mfile
function [x,numiter]=conj_grad(A,b,x) % Output x and inputs A and b
n=length(b); % Length must be same
b=b'; % Column vector
x=x'; % Initial guess will be column vector
r=b-A*x; % Column vector
 s=r;
p=s'; % Row vector
for numiter =1:n
  t=A*s;
alpha=p*r/(p*(t)); % alpha is defined
x=x+alpha*s; %x(k+1)=x(k)+alpha*s
r=b-A*x;
p=r'; % Row vector
beta1=A*s;
beta=-p*beta1/dot(s,beta1); % beta is defined
s=r+beta*s; %s(k+1)=r(k)+beta*s
end
```

```
>> % Command Window
>> conj_grad([10 -5 -2;4 -10 3;1 6 10],[1 -1 -1],[0 0 0])
ans =
   0.1313
  -0.0649
  -0.1514
```

2.7.4 CONVERGENCE

The convergence criterion of iterative methods is achieved when the absolute difference between the two consecutive iterations is less than some small quantity ϵ, i.e.,

$$\left| x^{j+1}_i - x^j_i \right| \leq \epsilon \text{ for } i = 1,2,\ldots n.$$

Or

$$\left| \frac{x^{j+1}_i - x^j_i}{x^j_i} \right| \leq \epsilon \text{ for } i = 1,2,\ldots n.$$

Problem Set 2

1. Given

$$2x_1 + x_2 + x_3 = 1,$$

(i) $3x_1 + 3x_2 + x_3 = 3,$

$$x_1 + 3x_2 + 2x_3 = 12.$$

$$x_1 + 2x_2 + x_3 + 2x_4 = 1,$$

$$x_1 + 8x_2 + x_3 + 3x_4 = 2,$$

(ii) $2x_1 + 6x_2 + x_3 + 9x_4 = 11,$

$$2x_1 + 8x_2 + 6x_3 + 2x_4 = 9.$$

$$3x_1 + 8x_2 + 2x_3 + 8x_4 = 11,$$
$$2x_1 + 11x_2 + 3x_3 + 6x_4 = -2,$$
(iii) $$2x_1 + 2x_2 + 9x_3 = 1,$$
$$-x_1 + 9x_2 + 16x_3 - 2x_4 = -9.$$

Use MATLAB to calculate the solutions of the given systems by using Cramer's rule.

2. Test your programs by using Gauss elimination and Gauss-Jordan methods to solve the following systems:

$$x_1 + 9x_2 + 2x_3 = 2,$$
(i) $$x_1 + 2x_2 + 2x_3 = 1,$$
$$x_1 + x_2 + 9x_3 = 2.$$

$$x_1 + 6x_2 + 2x_3 + 3x_4 = 1,$$
$$3x_1 + 3x_2 + 2x_3 + x_4 = 2,$$
(ii) $$2x_1 + x_2 + 3x_3 + x_4 = 2,$$
$$5x_1 + 3x_2 + 2x_3 + x_4 = 4.$$

$$x_1 + 2x_2 + x_3 + 9x_4 = 1,$$
$$2x_1 + 2x_2 + x_3 + 9x_4 = 3,$$
(iii) $$x_1 + 7x_2 + 2x_3 + x_4 = 1,$$
$$2x_1 + x_2 + 8x_3 + x_4 = 9.$$

3. Assume the following electrical network (Figure 2.2).

By using Kirchhoff's and Ohm's law we will get the following system:

$$3v_1 - 5v_2 + 2v_3 = 0,$$
$$-2v_1 + 5v_2 + 2v_4 = -5V,$$
$$v_2 - 3v_3 + 2v_4 = 0,$$
$$3v_1 + 10v_3 - 28v_4 = 0.$$

Where V is the potential between A and B and v_1, v_2, v_3 and v_4 denote the potentials. By using Doolittle's, Crout's and Choleski's methods, solve the problem for V = 20 and 40.

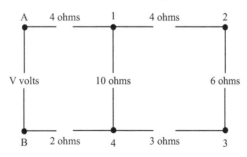

FIGURE 2.2 Electrical network.

4. By using MATLAB solve the systems by using Doolittle's, Crout's and Choleski's methods.

$$-x_1 + x_2 + 3x_3 = 12,$$
(i) $$x_1 + 9x_2 + 4x_3 = 11,$$
$$2x_1 + 3x_2 + 2x_3 = 9.$$

(ii)
$$4x_1 + 2x_2 + 3x_3 + 3x_4 = 2,$$
$$x_1 + 2x_2 + x_3 + 8x_4 = 1,$$
$$7x_1 + 2x_2 + x_3 + 2x_4 = 0,$$
$$5x_1 - 2x_2 + 6x_3 + 2x_4 = 4.$$

(iii)
$$x_1 - 2x_2 + 2x_3 + 2x_4 = 2,$$
$$x_1 - 2x_2 + 2x_3 + 3x_4 = 2,$$
$$x_1 + 2x_2 + x_3 - x_4 = 3,$$
$$x_1 - x_2 + 3x_3 - x_4 = 2.$$

5. Find L and U such that:

$$A = \begin{bmatrix} -4 & 1 & 0 \\ 1 & 4 & -1 \\ 0 & 1 & 4 \end{bmatrix}.$$

By using Doolittle's method.

6. Build a code of the following system $Ax = b$, where

$$A = \begin{bmatrix} 1 & 2 & 0 \\ 1 & 4 & 1 \\ 0 & 1 & 2 \end{bmatrix}, b = \begin{bmatrix} 2 \\ 3 \\ -1 \end{bmatrix},$$

by using Crout's and Choleski's methods.

7. Using MATLAB, solve the following tridiagonal system

$$\begin{bmatrix} 4 & 1 & 0 \\ 3 & 4 & 1 \\ 0 & 3 & 4 \end{bmatrix} \begin{bmatrix} x_1 \\ x_2 \\ x_3 \end{bmatrix} = \begin{bmatrix} -1 \\ 2 \\ 3 \end{bmatrix},$$

by using banded Doolittle's and Crout's methods.

8. Solve the system

$$A = \begin{bmatrix} 3 & 1 & 0 & 0 \\ -1 & 3 & 1 & 0 \\ 0 & -1 & 3 & 1 \\ 0 & 0 & -1 & 3 \end{bmatrix} \text{ and } b = \begin{bmatrix} -8 \\ -8 \\ -8 \\ -8 \end{bmatrix},$$

by using banded Crout's and Choleski's methods.

9. In the force formulation of a truss (see Figure 2.3), the unknown forces are v_i. For the statically determinate truss, the equilibrium equations of the joints are:

$$\begin{bmatrix} -1 & 1 & -1/\sqrt{2} & 0 & 0 & 0 \\ 0 & 0 & 1/\sqrt{2} & 1 & 0 & 0 \\ 0 & -1 & 0 & 0 & -1/\sqrt{2} & 0 \\ 0 & 0 & 0 & 0 & 1/\sqrt{2} & 0 \\ 0 & 0 & 0 & 0 & 1/\sqrt{2} & 1 \\ 0 & 0 & 0 & -1 & -1/\sqrt{2} & 0 \end{bmatrix} \begin{bmatrix} v_1 \\ v_2 \\ v_3 \\ v_4 \\ v_5 \\ v_6 \end{bmatrix} = \begin{bmatrix} 0 \\ 18 \\ 0 \\ 12 \\ 0 \\ 0 \end{bmatrix}.$$

Solve the system by using MATLAB.

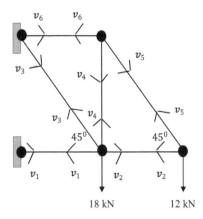

FIGURE 2.3 Forces on a truss.

10. Solve the following tridiagonal system by using MATLAB

$$A = \begin{bmatrix} \begin{bmatrix} -2 & 0 \\ 0 & 1 \end{bmatrix} & \begin{bmatrix} 0 & 0 \\ 0 & 0 \end{bmatrix} & \begin{bmatrix} 0 & 0 \\ 0 & 0 \end{bmatrix} \\ \begin{bmatrix} 1 & 0 \\ 0 & 1 \end{bmatrix} & \begin{bmatrix} -1 & -0.1 \\ -0.1 & 2 \end{bmatrix} & \begin{bmatrix} 1 & 0 \\ 0 & 1 \end{bmatrix} \\ \begin{bmatrix} 0 & 0 \\ 0 & 0 \end{bmatrix} & \begin{bmatrix} 1 & 0 \\ 0 & 1 \end{bmatrix} & \begin{bmatrix} 2 & 0 \\ 0 & 1 \end{bmatrix} \end{bmatrix}, R = \begin{bmatrix} \begin{bmatrix} 1 \\ 1 \end{bmatrix} \\ \begin{bmatrix} 0 \\ 0 \end{bmatrix} \\ \begin{bmatrix} -1 \\ 0 \end{bmatrix} \end{bmatrix}.$$

11. By using MATLAB programming devolve a code for the following system

$$A = \begin{bmatrix} \begin{bmatrix} 1 & 0 \\ 0 & 2 \end{bmatrix} & \begin{bmatrix} 0 & 0 \\ 0 & 0 \end{bmatrix} & \begin{bmatrix} 0 & 0 \\ 0 & 0 \end{bmatrix} \\ \begin{bmatrix} 3 & 0 \\ 0 & 3 \end{bmatrix} & \begin{bmatrix} 1 & 0.01 \\ -1 & 2 \end{bmatrix} & \begin{bmatrix} 1 & 0 \\ 0 & 1 \end{bmatrix} \\ \begin{bmatrix} 0 & 0 \\ 0 & 0 \end{bmatrix} & \begin{bmatrix} 1 & 2 \\ 0 & 1 \end{bmatrix} & \begin{bmatrix} 3 & 1 \\ 0 & 1 \end{bmatrix} \end{bmatrix} \begin{bmatrix} \begin{bmatrix} x_1 \\ y_1 \end{bmatrix} \\ \begin{bmatrix} x_2 \\ y_2 \end{bmatrix} \\ \begin{bmatrix} x_3 \\ y_3 \end{bmatrix} \end{bmatrix}, R = \begin{bmatrix} \begin{bmatrix} 0 \\ 1 \end{bmatrix} \\ \begin{bmatrix} 0 \\ 0 \end{bmatrix} \\ \begin{bmatrix} -1 \\ 1 \end{bmatrix} \end{bmatrix}.$$

12. Develop a code for the system

$$7x_1 - 2x_2 + 3x_3 = 16,$$
$$x_1 - 9x_2 + x_3 + x_4 = 12,$$
$$2x_1 + 10x_3 + 2x_4 = 10,$$
$$x_1 - x_2 + x_3 + 6x_4 = 4.$$

By using the Gauss-Jacobi method.

13. Using MATLAB, solve the following system:

$$\begin{bmatrix} 3 & 1 & 1 \\ -2 & 4 & 1 \\ -1 & 2 & -6 \end{bmatrix} \begin{bmatrix} x_1 \\ x_2 \\ x_3 \end{bmatrix} = \begin{bmatrix} 4 \\ 1 \\ 3 \end{bmatrix}.$$

By using the Gauss-Jacobi method.

14. Given the linear system

$$\begin{bmatrix} 4 & 0 & 1 \\ 1 & 4 & 1 \\ -1 & 0 & 4 \end{bmatrix} \begin{bmatrix} x_1 \\ x_2 \\ x_3 \end{bmatrix} = \begin{bmatrix} 5 \\ 3 \\ 10 \end{bmatrix}.$$

Use Gauss-Seidel method to calculate the solution by using MATLAB.

15. Solve the following system:

$$x_1 - x_2 + 2x_3 - x_4 = -1,$$
$$2x_1 + x_2 - 2x_3 - 2x_4 = -2,$$
$$-x_1 + 2x_2 - 4x_3 + x_4 = 1,$$
$$3x_1 - 2x_4 = -3.$$

by using the Gauss-Seidel method.

16. Solve the system

$$5x_1 - x_2 + 3x_3 = 3,$$
$$4x_1 + 7x_2 - 4x_3 = 2,$$
$$6x_1 - 3x_2 + 2x_3 = 9.$$

by using the conjugate gradient method.

17. Use the conjugate gradient method to solve

$$\begin{bmatrix} 3 & 0 & -1 \\ 0 & 4 & 2 \\ -1 & -3 & 5 \end{bmatrix} \begin{bmatrix} x \\ y \\ z \end{bmatrix} = \begin{bmatrix} 4 \\ 10 \\ -10 \end{bmatrix}.$$

18. Test your programs by using the conjugate gradient method to find the solution of the following system:

$$A = \begin{bmatrix} 1.2 & 0.45 & 0.35 & 0.45 \\ 0.89 & 2.59 & -0.33 & -0.22 \\ 0.72 & 0.77 & 4.02 & -0.88 \end{bmatrix} \text{ and } b = \begin{bmatrix} 2.61 \\ -11.78 \\ -15.03 \end{bmatrix}.$$

3 Polynomial Interpolation

Polynomial interpolation is used to calculate the intermediate values between the given data points. For example: The population data of Pakistan is taken from the previous six years. The population data table is shown in Table 3.1.

From the above table, the intermediate data, say, of 2017, can be predicated. There are various numerical methods that can be used to interpolate a polynomial from a given data. In this chapter those methods are discussed in detail with MATLAB® codes.

A mapping from a set of real numbers into itself is called algebraic mapping. The set of functions have the form:

$$P_n(x) = a_n x^{n-1} + a_{n-1} x^{n-2} \ldots\ldots a_2 x + a_1. \tag{3.1}$$

TABLE 3.1

Table of Population Data

Year	2012	2014	2016	2018	2020	2022
Population	202,205,861	208,251,628	213,524,840	216,379,655	227,196,741	235,824,862

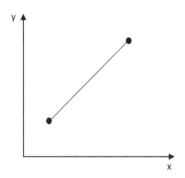

FIGURE 3.1 (a) First order interpolating polynomial.

Where $a_n, a_{n-1}, \ldots a_1$ are real constants and n is a nonnegative integer. Such types of polynomials uniformly approximate the continuous functions. This implies that for any given function which is continuous, defined on a bounded and closed interval contains a polynomial which is close to the given function. For n data points, there is only one polynomial of order $(n-1)$ that passes through all points. For example: Straight line is the only first order polynomial that joins the two points (as shown in Figure 3.1 a). Similarly, there is only one parabola (second order polynomial) that joins the three points (as shown in Figure 3.1 b).

The main reason that polynomial interpolation is mostly preferred is due to computations for definite values, differentiation and integration of such functions which are easy to calculate. Assume that if a function $f(x)$ is continuous in a close interval $[x_0, x_n]$, then for $\epsilon > 0$ there exists a polynomial $p(x)$ such that $|f(x) - p(x)| < \epsilon$. This implies that we can find a polynomial $p(x)$ such that within the bounded region ($y = f(x) - \epsilon$ and $y = f(x) + \epsilon$) the graph remains similar.

DOI: 10.1201/9781003385288-3

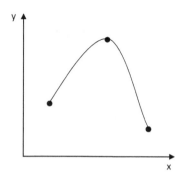

FIGURE 3.1   (b) Second order interpolating polynomial.

3.1   ERRORS IN POLYNOMIAL INTERPOLATION

Assume a function $y(x)$ be continuous and differentiable for n + 1 points, and assume that the function $y(x)$ is approximated by a polynomial $p_n(x)$ such that

$p_n(x_i) = y_i$ where $i = 0,1,2,...n$. As the expression $y(x) - p_n(x)$ approaches to zero for $x = x_0, x_1,...x_n$, so we substitute

$$y(x) - p_n(x) = R(x - x_0)(x - x_1)...(x - x_n).$$ (3.2)

As R is calculated for any intermediate value of x, say $\tilde{x}$, $x_0 < \tilde{x} < x_n$. Hence

$$R = \frac{y(\tilde{x}) - p_n(\tilde{x})}{(\tilde{x} - x_0)(\tilde{x} - x_1)...(\tilde{x} - x_n)}.$$ (3.3)

Construct a function $g(x)$ such that

$$g(x) = y(x) - p_n(\tilde{x}) - R(x - x_0)(x - x_1)...(x - x_n).$$ (3.4)

It is clear that

$$g(x_0) = g(x_1) = ... = g(x_n) = g(\tilde{x}) = 0.$$

Since $g(x)$ vanishes n + 2 times in the interval $[x_0, x_n]$. By Rolle's theorem, $g'(x)$ vanishes n + 1 times, $g''(x)$ vanishes n times, etc., in a similar manner $g^{n+1}(x)$ vanishes in the interval $[x_0, x_n]$. Denote the point by $x = \chi$, where $x_0 < \chi < x_n$. Differentiate Eq. (3.4) n + 1 times with respect to x and put $x = \chi$, we will get

$$0 = y^{n+1}(\chi) - R(n+1)!,$$

or

$$R = \frac{y^{n+1}(\chi)}{(n+1)!}.$$ (3.5)

Comparing Eqs. (3.3) and (3.5) yields

$$y(\tilde{x}) - p_n(\tilde{x}) = \frac{y^{n+1}(\chi)}{(n+1)!}(\tilde{x} - x_0)(\tilde{x} - x_1)...(\tilde{x} - x_n),$$

dropping the sign ~ gives:

$$y(x) - p_n(x) = \frac{y^{n+1}(\chi)}{(n+1)!}(x - x_0)(x - x_1)...(x - x_n), \text{where } x_0 < \chi < x_n. \quad (3.6)$$

Eq. (3.6) is the expression for errors in polynomials.

3.2 NEWTON'S FORWARD INTERPOLATION FORMULA (FOR EQUAL INTERVALS)

Let $x_0, x_1,...,x_n$ be required inputs with outputs achieved in the form $y_0, y_1,...,y_n$. Assume that the input points are equidistant from each other, i.e., $x_i - x_{i-1} = h$ where i = 1,2,...,n. Or $x_i = x_0 + ih$.

Assume $P_n(x)$ be an nth degree polynomial such that $y_i = f(x_i) = P_n(x)$.

$$P_n(x) = a_0 + a_1(x - x_0) + a_2(x - x_0)(x - x_1)$$
$$+...a_n(x - x_0)(x - x_1)...(x - x_{n-1}). \quad (3.7)$$

The unknowns $a_0, a_1...$ are calculated. First put $x = x_0$ in Eq. (3.7) and we have:

$$P_n(x_0) = y_0 = a_0.$$

For remaining terms use the relation: $\Delta^m P_n(x) = a_m m! h^m + \text{terms having} (x - x_0)$.

Put $x = x_0$ and rearranging gives $a_m = \frac{\Delta^m y_0}{m! h^m}$ where m = 1,2...n (mentioned in [2–30]). Hence substituting these values in Eq. (3.7) gives:

$$P_n(x) = y_0 + u\Delta\ y_0 + \frac{u(u-1)}{2!}\Delta^2 y_0$$
$$+...\frac{u(u-1)(u-2)...(u-n+1)}{n!}\Delta^n y_0 \quad (3.8)$$

The required polynomial (Eq. (3.8)) is known as Newton's forward interpolation formula. Here the symbol Δ denotes an operation known as forward difference operation and u = $\frac{x - x_0}{h}$.

The forward difference is defined as

$$\Delta y_0 = y_1 - y_0,$$

$$\Delta^2 y_0 = \Delta y_1 - \Delta y_0,$$

$$...$$

$$\Delta^n y_0 = \Delta^{n-1} y_1 - \Delta^{n-1} y_0. \quad (3.9)$$

In tabular form for x_k and y_k, where k = 0, 1,...,6, we have the information shown in Table 3.2.

This table is known as the diagonal difference table. The first value y_0 is called the leading term and the differences $\Delta\ y_0, \Delta^2 y_0, ...$ are leading differences.

To calculate the error in Newton's forward interpolation formula, use Eq. (3.6), i.e.,

$$y(x) - p_n(x) = \frac{y^{n+1}(\chi)}{(n+1)!}(x - x_0)(x - x_1)...(x - x_n), \text{where } x_0 < \chi < x_n. \quad (3.10)$$

TABLE 3.2

Forward Difference Table for k = 0, 1,…,6

x	y	Δy	$\Delta^2 y$	$\Delta^3 y$	$\Delta^4 y$	$\Delta^5 y$	$\Delta^6 y$
x_0	y_0						
x_1	y_1	Δy_0					
x_2	y_2	Δy_1	$\Delta^2 y_0$				
x_3	y_3	Δy_2	$\Delta^2 y_1$	$\Delta^3 y_0$			
x_4	y_4	Δy_3	$\Delta^2 y_2$	$\Delta^3 y_1$	$\Delta^4 y_0$		
x_5	y_5	Δy_4	$\Delta^2 y_3$	$\Delta^3 y_2$	$\Delta^4 y_1$	$\Delta^5 y_0$	
x_6	y_6	Δy_5	$\Delta^2 y_4$	$\Delta^3 y_3$	$\Delta^4 y_2$	$\Delta^5 y_1$	$\Delta^6 y_0$

As there is no information of $y^{n+1}(\chi)$, so a more useful result is achieved from this one. By using Taylor's series and neglecting the terms higher than h^2, we will get

$$y'(x) = \frac{1}{h}\left(y(x+h) - y(x)\right) = \frac{1}{h}\Delta y(x).$$

Write $y'(x)$ as $Dy(x)$ where $D = \dfrac{d}{dx}$ is the differentiation operator. Hence the above equation can be written as $D = \dfrac{1}{h}\Delta$ which implies that $D^{n+1} = \dfrac{1}{h^{n+1}}\Delta^{n+1}$. Thus

$$y^{n+1}(x) = \frac{1}{h^{n+1}}\Delta^{n+1}y(x). \tag{3.11}$$

Hence Eq. (3.10) becomes

$$y(x) - p_n(x) = \frac{\Delta^{n+1}y(\chi)}{(n+1)!}u(u-1)(u-2)\ldots(u-n), \text{where } x_0 < \chi < x_n. \tag{3.12}$$

Which is the more suitable form.

Example 3.1: Find the values of y at x = 13 from the following data:

x	12	14	16	18
y	0.3220	0.3856	0.4456	0.4997

Where h = 2.

Solution: First construct a table:

x	y	Δy	$\Delta^2 y$	$\Delta^3 y$
12	0.3220			
14	0.3856	0.0636		
16	0.4456	0.0600	−0.0036	
18	0.4997	0.0541	−0.0059	−0.0023

For Newton's forward interpolation formula diagonal entries are needed. By formula

$$P_3(x) = y_0 + u\Delta y_0 + \frac{u(u-1)}{2!}\Delta^2 y_0 + \frac{u(u-1)(u-2)}{3!}\Delta^3 y_0.$$

Where $u = \dfrac{x - x_0}{h} = \dfrac{13 - 12}{2} = 0.500$. Hence

$$P_3(13) = 0.3220 + (0.500)(0.0636)$$

$$+ \frac{0.500(0.500 - 1)}{2!}(-0.0036)$$

$$+ \frac{0.500(0.500 - 1)(0.500 - 2)}{3!}(-0.0023),$$

$$P_3(13) = 0.3541.$$

Main steps for solving Newton's forward interpolation formula in MATLAB:

1) The inputs and outputs are defined first. As x and y are given data and t is the value which is to be calculated.
2) Length of x is n, i.e., n = length (x).
3) A matrix n × n is created, i.e., A = zeros(n,n).
4) Next u is defined by xx = (t − x(1))/h.
5) From table y is a column vector, i.e., A(:,1) = y'.
6) A loop is generated to get the required table in such a way that second entry minus first entry, third entry minus second entry and vice versa. For this, two loops are implemented in such a way A(i,j) =A (i,j − 1)-A(i − 1,j − 1) where the first loop starts from j = 2 to n + 1 and the second loop starts from i = j to n. From this operation the first column is operated, then second and so on.
7) Denote y0 = A(1,1).
8) Now for creating a forward difference polynomial in MATLAB, define

xt = 1 (create a sequence),

s = 0 (add the sequence),

xt = xt * (xx − j) (creating a sequence),

xm = xt/factorial(j + 1) (sequence created),

S = s + xm * A(j + 2,j + 2) (addition of sequence).

        Note: The loop will start from 0 to n − 2.

9) At last, the y0 term and sum terms are added to get the required result.

```
% Mfile
function [p,A]= Newton_forward_differnce(x,y,t)
% Inputs and outputs are defined.
% x and y are required data
% t is value which is to be calculated
n=length(x);
A=zeros(n,n); % Consider a n*n matrix
h=2;
xx=(t-x(1))/h; % By formula xx is u
A(:,1)=y'; % As the given values of y are column vector
for j=2:n+1 % The remaining table is calculated by 2nd-1st.
  for i=j:n
    A(i,j)=A(i,j-1)-A(i-1,j-1);
  end
end
% For Newton's forward interpolation formula
% Diagonal entries are needed
```

```
xt=1;


y0=A(1,1); % By formula 1st entry of y is y0
s=0;
for j=0:n-2
 xt=xt*(xx-j); % A loop is generated for required formula
 xm=xt/factorial(j+1);
 s=s+xm*A(j+2,j+2);
end
p=y0+s;
```

```
>> % Command Window
>> Newton_forward_differnce([12 14 16 18],[0.3220 0.3856 0.4456
0.4997],13)
ans =
  0.3541
```

3.3 NEWTON'S BACKWARD INTERPOLATION FORMULA (FOR EQUAL INTERVALS)

Near the end points of the table, Newton's forward interpolation formula cannot be used. For this reason, another interpolating method known as Newton's backward interpolation is used. Let x_0, $x_1,...,x_n$ be required inputs with outputs achieved in the form y_0, $y_1,...,y_n$. Assume that the input points are equidistant from each other, i.e., $x_i - x_{i-1} = h$ or $x_i = x_0 + ih$, where i = 1,2,.....n.

Assume $P_n(x)$ be an nth degree polynomial such that $y_i = f(x_i) = P_n(x)$. By symbolic operator method

$$P_n(x) = P_n(x_n + qh) = E^q P_n(x_n) = (1 - \nabla)^{-q} y_n,$$

as $E = (1 - \nabla)^{-1}$ (mentioned in [2–30]).
   Expanding gives:

$$P_n(x) = y_n + q\nabla y_n + \frac{q(q+1)}{2!} \nabla^2 y_n + ...$$

$$\frac{q(q+1)(q+2)...(q+n-1)}{n!} \nabla^n y_n. \tag{3.13}$$

The required polynomial is known as Newton's backward interpolation formula. Here the symbol ∇ denotes an operation known as backward difference operation and q = $\dfrac{x - x_n}{h}$.

   In tabular form for x_k and y_k where k = 0, 1,...,6, we have the information shown in Table 3.3.
   The final row is used in Newton's backward interpolation formula. The error of this formula is written as

$$y(x) - p_n(x) = \frac{\nabla^{n+1} y(\chi)}{(n+1)!} q(q+1)(q+2)...(q+n), \text{ where } x_0 < \chi < x_n. \tag{3.14}$$

TABLE 3.3
Backward Difference Table for k = 0, 1,...,6

x	y	∇y	$\nabla^2 y$	$\nabla^3 y$	$\nabla^4 y$	$\nabla^5 y$	$\nabla^6 y$
x_0	y_0						
x_1	y_1	∇y_0					
x_2	y_2	∇y_1	$\nabla^2 y_0$				
x_3	y_3	∇y_2	$\nabla^2 y_1$	$\nabla^3 y_0$			
x_4	y_4	∇y_3	$\nabla^2 y_2$	$\nabla^3 y_1$	$\nabla^4 y_0$		
x_5	y_5	∇y_4	$\nabla^2 y_3$	$\nabla^3 y_2$	$\nabla^4 y_1$	$\nabla^5 y_0$	
x_6	y_6	∇y_5	$\nabla^2 y_4$	$\nabla^3 y_3$	$\nabla^4 y_2$	$\nabla^5 y_1$	$\nabla^6 y_0$

Example 3.2: Find the values of y at x = 17 from the following data:

x	12	14	16	18
y	0.3220	0.3856	0.4456	0.4997

Where h = 2.

Solution: First construct a table:

x	y	Δy	$\Delta^2 y$	$\Delta^3 y$
12	0.3220			
14	0.3856	0.0636		
16	0.4456	0.0600	−0.0036	
18	0.4997	0.0541	−0.0059	−0.0023

As can be seen from the above table, for Newton's backward interpolation formula final entries are needed. By formula

$$P_3(x) = y_n + q\nabla y_n + \frac{q(q+1)}{2!}\nabla^2 y_n + \frac{q(q+1)(q+2)}{3!}\nabla^3 y_n.$$

Where q = $\dfrac{x - x_n}{h} = \dfrac{17 - 18}{2} = -0.500$. Hence

$$P_3(17) = 0.4997 + (-0.500)(0.0541)$$

$$+ \frac{-0.500(-0.500+1)}{2!}(-0.0059)$$

$$+ \frac{-0.500(-0.500+1)(-0.500+2)}{3!}(-0.0023).$$

$$P_3(17) = 0.4735.$$

Main steps for solving Newton's backward interpolation formula in MATLAB:

1) The inputs and outputs are defined first. As x, y and t are given.
2) Length of x is n, i.e., n = length (x).

3) A matrix n × n is created.
4) Next q is defined by xx =(t − x(n))/h.
5) From table y is a column vector, i.e., A(:,1) = y′.
6) A loop is generated to get the required table in such a way that second entry minus first entry, third entry minus second entry and vice versa.
7) Denote yn = A(n,1) (nth row, first entry).
8) Now for creating a backward difference polynomial in MATLAB, define

xt = 1 (create a sequence),

s = 0 (add the sequence),

The loop will start from 0 to n − 2,

xt = xt * (xx + j) (creating a sequence),

xm = xt/factorial(j + 1) (sequence created),

s =s + xm * A(n,j +2) (addition of sequence).

9) At the end, the yn term and s are added.

```
% Mfile
function [p,A]= Newton_backward_differnce(x,y,t)
% Inputs and outputs are defined.
% x, y and t are given data
% t is value which is to be calculated
n=length(x);
A=zeros(n,n); % Consider a n*n matrix
h=2;
xx=(t-x(n))/h; % By formula xx is q
A(:,1)=y'; % As the given values of y is a column vector
for j=2:n+1 % The remaining table is calculated by 2nd-1st
    for i=j:n
     A(i,j)=A(i,j-1)-A(i-1,j-1);
   end
end
% For Newton's backward interpolation formula
yn=A(n,1); % nth entry of the 1st column
xt=1;
s=0;
for j=0:n-2
 xt=xt*(xx+j);
 xm=xt/factorial(j+1);
s=s+xm*A(n,j+2);
   end
p=yn+s;
```

```
   >> % Command Window
>> Newton_backward_differnce([12 14 16 18],[0.3220 0.3856 0.4456
0.4997],17)
ans =
  0.4735
```

3.4   NEWTON'S DIVIDED DIFFERENCE INTERPOLATION (UNEQUAL INTERVALS)

For Newton's divided difference, we choose a polynomial interpolating at $x = x_i$, where i = 1,2..n, such that

$$f(x) = P_n(x) = a_0 + a_1(x - x_0) + a_2(x - x_0)(x - x_1)$$

$$+ \ldots u_n(x - x_0)(x - x_1)\ldots(x - x_{n-1}).$$

(3.15)

The coefficients are calculated from Eq. (3.15). Thus

$$f(x_0) = a_0,$$

$$f(x_1) = a_0 + a_1(x_1 - x_0),$$

(3.16)

$$f(x_2) = a_0 + a_1(x_2 - x_0) + a_2(x_2 - x_0)(x_2 - x_1).$$

Calculating the values of a_1 and a_2 from Eq. (3.16), we have:

$$a_1 = \frac{f(x_1) - f(x_0)}{x_1 - x_0} = f[x_0, x_1], \text{ (second divided difference)}$$

in a similar manner,

$$a_2 = \frac{f[x_1, x_2] - f[x_0, x_1]}{x_2 - x_0} = f[x_0, x_1, x_2], \text{ (third divided difference)}.$$

Hence Newton's divided difference formula can be written as

$$f(x) = f(x_0) + (x - x_0)f[x_0, x_1]$$

$$+ (x - x_0)(x - x_1)f[x_0, x_1, x_2]$$

(3.17)

$$+ \ldots(x - x_0)(x - x_1)\ldots(x - x_{n-1})f[x_0, x_1, x_2 \ldots x_{n-1}, x_n],$$

since

$$f[x_0, x_1] = \frac{f(x_1) - f(x_0)}{x_1 - x_0} = \frac{\Delta f(x_0)}{h},$$

in a similar way,

$$f[x_0, x_1, x_2] = \frac{f[x_1, x_2] - f[x_0, x_1]}{x_2 - x_0}$$

$$= \frac{\dfrac{\Delta f(x_1)}{h} - \dfrac{\Delta f(x_0)}{h}}{2h} = \frac{\Delta^2 f(x_0)}{h^2 2!}.$$

The divided difference table for this procedure is mentioned in Table 3.4.

The error in divided difference polynomial is calculated from Eq. (3.6), which is

$$y(x) - p_n(x) = \frac{y^{n+1}(\chi)}{(n+1)!}(x - x_0)(x - x_1)\ldots(x - x_n),$$

where $x_0 < \chi < x_n$.

TABLE 3.4

Divided Difference Table

X	$f(x_i)$	$f[x_i,x_{i+1}]$	$f[x_i,x_{i+1},x_{i+2}]$	$f[x_i,x_{i+1},x_{i+2},x_{i+3}]$	$f[x_i,x_{i+1},x_{i+2},x_{i+3},x_{i+4}]$
x_1	$y_1 = A(1,1)$				
x_2	$y_2 = A(2,1)$	$A(2,2) = A(2,1)$ $-A(1,1)/(x_2-x_1)$			
x_3	$y_3 = A(3,1)$	$A(3,2) = A(3,1)$ $-A(2,1)/(x_3-x_2)$	$A(3,3) = A(3,2)$ $-A(2,2)/(x_3-x_1)$		
x_4	$y_4 = A(4,1)$	$A(4,2) = A(4,1)$ $-A(3,1)/(x_4-x_3)$	$A(4,3) = A(4,2)$ $-A(3,2)/(x_4-x_2)$	$A(4,4) = A(4,3)$ $-A(3,3)/(x_4-x_1)$	
x_5	$y_5 = A(5,1)$	$A(5,2) = A(5,1)$ $-A(4,1)/(x_5-x_4)$	$A(5,3) = A(5,2)$ $-A(4,2)/(x_5-x_3)$	$A(5,4) = A(5,3)$ $-A(4,3)/(x_5-x_2)$	$A(5,5) = A(5,4) - A(4,4)/$ (x_5-x_1)

As the function and its derivatives are unknown so replacing the $(n+1)$ *th* derivative with divided difference expression

$$y(x) - p_n(x) = f[x_0,x_1,x_2\ldots x_{n.}]$$
$$(x-x_0)(x-x_1)\ldots(x-x_n).$$

(3.18)

Example 3.3: Use Newton's divided difference formula to find the value of f(9) from a given table:

x	8	10	12	13
y	47	99	295	453

Solution: The divided difference table for the given data is:

x	f(x)	$f[x_1,x_2]$	$f[x_1,x_2,x_3]$	$f[x_1,x_2,x_3,x_4]$
8	47			
10	99	(99 − 47)/(10 − 8) = 26		
12	295	(295 − 99)/(12 − 10) = 98	(98 − 26)/(12 − 8) = 18	
13	453	(453 − 295)/(13 − 12) = 158	(158 − 98)/(13 − 10) = 20	(20 − 18)/(13 − 8) = 0.4

By formula

$$f(9) = 47 + (9-8)26 + (9-8)(9-10)\ 18 + (9-8)(9-10)(9-12)0.4 = 56.2.$$

Main steps for solving the Newton's divided difference interpolation formula in MATLAB:

1) The procedure is the same from the inputs/outputs to the column vector.
2) A loop is generated to get the required table in such a way that second entry minus first entry, third entry minus second entry and vice versa divided by (x(i) − x(i − j + 1)), where the outer loop starts from j = 2 to n and inner loop starts from i = j to n.
3) For Newton's divided interpolation a sequence is created and added with a loop from 1 to n − 1.

4) At last, the first term of y and the sequence (s) is added.

```
% Mfile
function [p,A]= Newton_divided_differnce(x,y,t)
% Inputs and outputs are defined.
% x and y are given data
% t is the value which is to be calculated
n=length(x);
A=zeros(n,n); % Consider a n*n matrix
A(:,1)=y'; % As the given values of y is the column vector
for j=2:n % The remaining table is calculated
  for i=j:n
    A(i,j)=(A(i,j-1)-A(i-1,j-1))/(x(i)-x(i-j+1));
  end
end
xt=1; % For creating a sequence
s=0; % Adding a sequence
y0=A(1,1); % First entry of y
for j=1:n-1
 xt=xt*(t-x(j)); % Sequence created
 s=s+A(j+1,j+1)*xt; % Addition of a sequence
end
p=y0+s; % Formula
```

```
>> % Command Window
>> Newton_divided_differnce([8 10 12 13],[47 99 295 453],9)
ans =
  56.2000
```

3.5   LAGRANGE'S INTERPOLATION FORMULA

The weighted average of two linear interpolating polynomials connected by a straight line is:

$$f(x) = l_0 f(x_0) + l_1 f(x_1). \tag{3.19}$$

Where l_0 and l_1 are weighted coefficients (as shown in Figure 3.2). The first weighted coefficient is a straight line that is equal to 1 at x_0 and 0 at x_1:

$$l_0 = \frac{x - x_1}{x_0 - x_1}. \tag{3.20}$$

In a similar way, the second coefficient is a straight line passing through 1 at x_1 and 0 at x_0:

$$l_1 = \frac{x - x_0}{x_1 - x_0}. \tag{3.21}$$

Putting the values gives:

$$f_0(x) = \frac{x - x_1}{x_0 - x_1} f(x_0) + \frac{x - x_0}{x_1 - x_0} f(x_1). \tag{3.22}$$

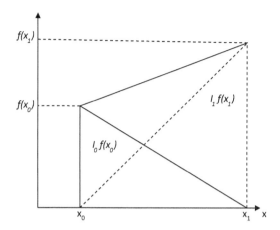

FIGURE 3.2 The Lagrange interpolating polynomial.

The same procedure is applied to fit a parabola through three points. Three parabolas are used with each parabola passing through one of the points and zero at other points. The sum represents a unique parabola that passes through three points. The second order Lagrange interpolation becomes:

$$f_1(x) = \frac{(x-x_1)(x-x_2)}{(x_0-x_1)(x_0-x_2)} f(x_0)$$

$$+ \frac{(x-x_0)(x-x_2)}{(x_1-x_0)(x_1-x_2)} f(x_1) \tag{3.23}$$

$$+ \frac{(x-x_0)(x-x_1)}{(x_2-x_0)(x_2-x_1)} f(x_2).$$

In a similar way, the generalized form is:

$$f(x) = \frac{(x-x_1)(x-x_2)...(x-x_n)}{(x_0-x_1)(x_0-x_2)...(x_0-x_n)} f(x_0)$$

$$+ \frac{(x-x_0)(x-x_2)...(x-x_n)}{(x_1-x_0)(x_1-x_2)...(x_1-x_n)} f(x_1) + ... \tag{3.24}$$

$$+ \frac{(x-x_0)(x-x_1)...(x-x_{n-1})}{(x_n-x_0)(x_n-x_2)...(x_n-x_{n-1})} f(x_n).$$

The above formula is called Lagrange's interpolation formula for unequal intervals.

Eq. (3.6) is used to calculate the error of Lagrange interpolating polynomial for the functions which have continuous derivatives up to n + 1 on [a,b], i.e.,

$$y(x) - p_n(x) = Q_n(x) = \frac{y^{n+1}(\chi)}{(n+1)!}(x-x_0)(x-x_1)...(x-x_n), \text{where} x_0 < \chi < x_n.$$

Define S_l as

$$S_l = \max_{[a,b]} |Q_n(x)|, \tag{3.25}$$

Eq. (3.25) is used to calculate the error. Moreover assume

$$|y^{n+1}(\chi)| \leq W_{n+1}, \text{ where } x_0 < \chi < x_n. \tag{3.26}$$

Then

$$S_l \leq \frac{W_{n+1}}{(n+1)!} \max_{[a,b]} |(x - x_0)(x - x_1)...(x - x_n)|. \tag{3.27}$$

Which is the required error.

Example 3.4: Using Lagrange's interpolation formula, calculate y(6) from the following data:

x	4	5	7
y	11	12	17

Solution: By using Lagrange's interpolation formula

$$f(x) = \frac{(x - x_1)(x - x_2)}{(x_0 - x_1)(x_0 - x_2)} f(x_0) + \frac{(x - x_0)(x - x_2)}{(x_1 - x_0)(x_1 - x_2)} f(x_1)$$

$$+ \frac{(x - x_0)(x - x_1)}{(x_2 - x_0)(x_2 - x_1)} f(x_2),$$

substituting the given values gives

$$f(x) = \frac{(x - 5)(x - 7)}{(4 - 5)(4 - 7)}(11) + \frac{(x - 4)(x - 7)}{(5 - 4)(5 - 7)}(12) + \frac{(x - 4)(x - 5)}{(7 - 4)(7 - 5)}(17),$$

for x = 6, we have:

$$f(6) = \frac{(6 - 5)(6 - 7)}{(4 - 5)(4 - 7)}(11) + \frac{(6 - 4)(6 - 7)}{(5 - 4)(5 - 7)}(12) + \frac{(6 - 4)(6 - 5)}{(7 - 4)(7 - 5)}(17),$$

$$f(6) = -3.6666 + 12 + 5.6666 = 14.$$

Main steps for solving Lagrange's interpolation formula in MATLAB:

1) The inputs and output are defined first. In inputs, x and y are given data and xx is the value which is going to be calculated.
2) Length of x is n. i.e., n = length (x).
3) For addition of sequence take t = 0.
4) A loop is created from j = 1 to n.
5) The given function is defined, i.e, p0 = y(i), where i = 1 to n.
6) A loop j starts from 1 to n.

7) By formula $i \neq j$, a condition 'if' is inserted to tackle this one.
8) The formula is written.
9) For addition of sequence (by (3)), t = t + p0 is written.
10) And the output is saved in the form of p, so put p = t.

```
% Mfile
function [p]= Lagrange_interpolation(x,y,xx)
% Inputs and outputs are defined.
% x and y are given data
% xx is the value which is to be calculated
n=length(x);
t=0; % Addition of sequence
for i=1:n
  p0=y(i); % Given data in the form of f
  for j=1:n
   if i~=j % By formula
  p0=p0*(xx-x(j))/(x(i)-x(j));
   end
  end
  t=t+p0; % Addition of sequence
end
p=t;
```

```
>> % Command Window
>> Lagrange_interpolation([4 5 7],[11 12 17],6)
```

3.6   NEVILLE'S METHOD

In Newton's method, coefficients are computed first and then the polynomial is evaluated. But in Neville's method, the whole procedure is performed in a single step. Let the function f be defined at $x_1, x_2,...,x_n$ then

$$A_{ij} = \frac{\left(x - x_{i-j+1}\right)A_{ij-1} - \left(x - x_i\right)A_{i-1j-1}}{x_i - x_{i-j+1}}. \tag{3.28}$$

Where i,j = 2,3...n, Table 3.5.

TABLE 3.5
Neville's Table for i and j = 2, 3, ...n

x	y = $f(x)$	A(i,2)	A(i,3)	A(i,4)	A(i,5)
x_1	A(1,1)				
x_2	A(2,1)	A(2,2)			
x_3	A(3,1)	A(3,2)	A(3,3)		
x_4	A(4,1)	A(4,2)	A(4,3)	A(4,4)	
x_5	A(5,1)	A(5,2)	A(5,3)	A(5,4)	A(5,5)

Example 3.5: By using Neville's method, construct a recursive table from given data:

x	1	1.2	1.4	1.6
y	0.6543	0.5876	0.4765	0.3654

And calculate the value at f(1.3).

Solution: By using formula:

$$A_{ij} = \frac{(x - x_{i-j+1})A_{ij-1} - (x - x_i)A_{i-1j-1}}{x_i - x_{i-j+1}}.$$

For i = 2, j = 2, we have

$$A_{22} = \frac{(x - x_1)A_{21} - (x - x_2)A_{11}}{x_2 - x_1} = \frac{(1.3-1)0.5876 - (1.3-1.2)0.6543}{1.2-1} = 0.5542.$$

For i = 3, j = 2, we have:

$$A_{32} = \frac{(x - x_2)A_{31} - (x - x_3)A_{21}}{x_3 - x_2} = \frac{(1.3-1.2)0.4765 - (1.3-1.4)0.5876}{1.4-1.2} = 0.5320.$$

For i = 4, j = 2, we have:

$$A_{32} = \frac{(x - x_3)A_{41} - (x - x_4)A_{31}}{x_4 - x_3} = \frac{(1.3-1.4)0.3654 - (1.3-1.6)0.4765}{1.4-1.2} = 0.5320.$$

Similarly,

$$A_{33} = 0.5376, A_{43} = 0.5320, A_{44} = 0.5348.$$

Main steps for solving Neville's method in MATLAB:

1) The inputs and output are defined first. In inputs, y and x are given data and yy is the value which is going to be calculated.
2) Length of y is n, i.e., n = length (y).
3) Consider a $n \times n$ matrix, i.e., zeros(n,n).
4) The values of y are taken as the first column vector of the matrix, i.e., A(:,1) = y'.
5) The remaining table is calculated by Neville's formula, i.e.,
   A(i,j) = ((yy – x(i – j + 1)) * A(i,j – 1) – (yy – x(i)) * A(i – 1,j – 1))/(x(i) – x(i – j + 1)), where the outer loop starts from j = 2 to n and the inner loop starts from i = j to n.

```
% Mfile
function [A]= Nevilles(y,x,yy)
% Inputs and output are defined.
% x and y are given data
% yy is the value which is to be calculated
n=length(y);
A=zeros(n,n); % Consider a n*n matrix
A(:,1)=y';% As the given values of y is a column vector
```

```
for j=2:n % The remaining table is calculated
% Using Nevilles formula
  for i=j:n
A(i,j)=((yy-x(i-j+1))*A(i,j-1)-(yy-x(i))*A(i-1,j-1))/(x(i)-x(i-j
+1));
  end
end
```

```
>> % Command Window
>> [A]=Nevilles([0.6543 0.5876 0.4765 0.3654],[1 1.2 1.4 1.6],1.3)
A =
   0.6543        0           0        0
   0.5876   0.5542           0        0
   0.4765   0.5320      0.5376        0
   0.3654   0.5320      0.5320   0.5348
```

3.7   CUBIC SPLINE INTERPOLATION

A cubic spline interpolation method uses cubic polynomials between each pair of nodes. It involves four constants and is continuously differentiable with a continuous second derivative on each interval. The cubic model is illustrated in Figure 3.3. The figure is divided into small segments, each segment of the curve is a cubic polynomial. At the data points, the second derivative is continuous. At the end points, there is no bending, hence the second derivative of the spline is zero.

Consider a function defined on [a,b] and the set of nodes $x_0 < x_1 < x_2 \ldots < x_n$ is said to be cubic spline interpolation M for function, if it satisfies the following conditions:

(i) $M_j(x_j) = f(x_j)$ for $j = 1, 2 \ldots n-1$.
(ii) $M_{j+1}(x_{j+1}) = M_j(x_{j+1})$ for $j = 1, 2 \ldots n-2$.
(iii) $M'_{j+1}(x_{j+1}) = M'_j(x_{j+1})$ for $j = 1, 2 \ldots n-2$.
(iv) $M''_{j+1}(x_{j+1}) = M''_j(x_{j+1})$ for $j = 1, 2 \ldots n-2$.
(v) $M''(x_0) = M''(x_n) = 0$ (free boundary conditions).

When the boundary conditions are free, then the spline is called a natural spline. The cubic polynomial is:

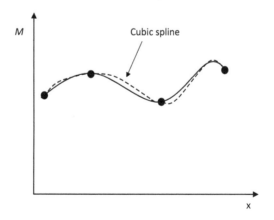

FIGURE 3.3   Cubic spline.

$$M_j(x) = A_j + B_j(x - x_j) + C_j(x - x_j)^2 + D_j(x - x_j)^3. \tag{3.29}$$

Where $j = 1, 2 \ldots n - 1$. By applying conditions (i) and (ii), we get:

$$A_{j+1} = A_j + B_j(x_{j+1} - x_j) + C_j(x_{j+1} - x_j)^2 + D_j(x_{j+1} - x_j)^3. \tag{3.30}$$

Where $j = 1, 2 \ldots n - 2$. Assume $h_j = x_{j+1} - x_j$ for $j = 1, 2 \ldots n - 1$.
 Define $B_n = M'(x_n)$. Hence:

$$M'_j(x) = B_j + 2C_j(x - x_j) + 3D_j(x - x_j)^2. \tag{3.31}$$

Since $M'_j(x_j) = B_j$ and applying condition (iii)

$$B_{j+1} = B_j + 2C_j h_j + 3D_j h_j^2. \tag{3.32}$$

Applying condition (iii) and defining $C_n = M''(x_n)/2$, we have:

$$C_{j+1} = C_j + 3D_j h_j \ . \tag{3.33}$$

Eliminating D_j:

$$D_j = \frac{(C_{j+1} - C_j)}{3h_j}. \tag{3.34}$$

By using (3.34) into Eqs. (3.30) and (3.32), we get:

$$A_{j+1} = A_j + B_j h_j + \frac{h_j^2}{3}(2C_j + C_{j+1}), \tag{3.35}$$

and

$$B_{j+1} = B_j + h_j(C_j + C_{j+1}). \tag{3.36}$$

Solving Eq. (3.35) for B_j, we get:

$$B_j = \frac{1}{h_j}(A_{j+1} - A_j) - \frac{h_j}{3}(2C_j + C_{j+1}). \tag{3.37}$$

Reducing the index for B_{j-1} gives:

$$B_{j-1} = \frac{1}{h_{j-1}}(A_j - A_{j-1}) - \frac{h_{j-1}}{3}(2C_{j-1} + C_j). \tag{3.38}$$

Putting these values in (3.36) and reducing the index by 1.

$$h_{j-1}C_{j-1} + 2(h_{j-1} + h_j)C_j + h_j C_{j+1} = \frac{3}{h_j}(A_{j+1} - A_j) - \frac{3}{h_{j-1}}(A_j - A_{j-1}). \tag{3.39}$$

Where $j = 1, 2 \ldots n - 1$. Here C_j, C_{j-1} and C_{j+1} are unknown and the remaining are given.

For natural splines $M''(a) = 0$ and $M''(b) = 0$.
Hence the matrix form becomes

$$
\begin{bmatrix}
1 & 0 & 0 & 0 & - & - & - & 0 \\
h_1 & 2(h_1 + h_2) & h_2 & & - & - & & 0 \\
0 & h_2 & 2(h_2 + h_3) & h_3 & - & - & & 0 \\
& - & - & - & - & - & & \\
0 & - & - & - & h_{n-2} & 2(h_{n-2} + h_{n-1}) & & h_{n-1} \\
0 & - & - & & & 0 & & 1
\end{bmatrix}
\begin{bmatrix}
C_1 \\
C_2 \\
- \\
- \\
C_n
\end{bmatrix}
=
$$

$$
\begin{bmatrix}
0 \\
\dfrac{3}{h_2}(A_3 - A_2) - \dfrac{3}{h_1}(A_2 - A_1) \\
- \\
\dfrac{3}{h_{n-1}}(A_n - A_{n-1}) - \dfrac{3}{h_{n-2}}(A_{n-1} - A_{n-2}) \\
0
\end{bmatrix}.
$$

(3.40)

The matrix will be diagonally dominant.

Example 3.6: Consider the data:

i	x_i	A_i
1	2.5	2
2	4	0.5
3	6.5	2
4	8.5	0.1

Use these results and estimate the value at x = 5.

Solution: First the values of coefficients C's are calculated from the vector

$$
\begin{bmatrix}
1 & 0 & 0 & 0 & - & - & - & 0 \\
h_1 & 2(h_1 + h_2) & h_2 & & - & - & & 0 \\
0 & h_2 & 2(h_2 + h_3) & h_3 & - & - & & 0 \\
& - & - & - & - & - & & \\
0 & - & - & - & h_{n-2} & 2(h_{n-2} + h_{n-1}) & & h_{n-1} \\
0 & - & - & & & 0 & & 1
\end{bmatrix}
\begin{bmatrix}
C_1 \\
C_2 \\
- \\
- \\
C_n
\end{bmatrix}
=
$$

$$
\begin{bmatrix}
0 \\
\dfrac{3}{h_2}(A_3 - A_2) - \dfrac{3}{h_1}(A_2 - A_1) \\
- \\
\dfrac{3}{h_{n-1}}(A_n - A_{n-1}) - \dfrac{3}{h_{n-2}}(A_{n-1} - A_{n-2}) \\
0
\end{bmatrix}.
$$

Here $A_1 = 2, A_2 = 0.5, A_3 = 2, A_4 = 0.1$.
And $h_1 = 4 - 2.5 = 1.5, h_2 = 6.5 - 4 = 2.5, h_3 = 8.5 - 6.5 = 2$.
Putting these results gives:

$$\begin{bmatrix} 1 & 0 & 0 & 0 \\ 1.5 & 8 & 2.5 & 0 \\ 0 & 2.5 & 9 & 2 \\ 0 & 0 & - & 1 \end{bmatrix} \begin{bmatrix} C_1 \\ C_2 \\ C_3 \\ C_4 \end{bmatrix} = \begin{bmatrix} 0 \\ 4.8 \\ -4.6 \\ 0 \end{bmatrix}.$$

To calculate the values of C's MATLAB is used which gives the result:

$$C_1 = 0, C_2 = 0.8338, C_3 = -0.7483, C_4 = 0.$$

Calculating B's and D's

$$B_1 = \frac{1}{h_1}(A_2 - A_1) - \frac{h_1}{3}(2C_1 + C_2) = -1.4169,$$

$$B_2 = \frac{1}{h_2}(A_3 - A_2) - \frac{h_2}{3}(2C_2 + C_3) = -0.1662,$$

$$B_3 = \frac{1}{h_3}(A_4 - A_3) - \frac{h_3}{3}(2C_3 + C_4) = 0.0477,$$

$$D_1 = \frac{(C_2 - C_1)}{3h_1} = 0.1853, \quad D_2 = \frac{(C_3 - C_2)}{3h_2} = -0.2110, \quad D_3 = \frac{(C_4 - C_3)}{3h_3} = 0.1247.$$

The calculated results are substituted into the cubic polynomial which gives cubic splines for each interval.

$$M_1(x) = A_1 + B_1(x - x_1) + C_1(x - x_1)^2 + D_1(x - x_1)^3,$$

$$M_1(x) = 2 - 1.4169(x - 2.5) + 0(x - 2.5)^2 + 0.1853(x - 2.5)^3,$$

$$M_2(x) = A_2 + B_2(x - x_2) + C_2(x - x_2)^2 + D_2(x - x_2)^3,$$

$$M_2(x) = 0.5 - 0.1662(x - 4) + 0.8338(x - 4)^2 + 0.2110(x - 4)^3,$$

$$M_3(x) = A_3 + B_3(x - x_3) + C_3(x - x_3)^2 + D_3(x - x_3)^3,$$

$$M_3(x) = 2 + 0.0477(x - 6.5) - 0.7483(x - 6.5)^2 + 0.1247(x - 6.5)^3.$$

These three equations are used to compute values within each interval. As the value x = 5 falls within the second interval so

$$M_2(5) = 0.5 - 0.1662(5 - 4) + 0.8338(5 - 4)^2 + 0.2110(5 - 4)^3 = 0.9567.$$

Main steps for solving the cubic spline interpolation in MATLAB:

1) The inputs and output are defined first. In inputs, x and f are given data and xx is the value which is going to be calculated.
2) Length of n will be b, i.e., n = length (b).
3) b is a column vector, i.e., b = b'.

4) f is a column vector, i.e., f = f'.
5) An n×n matrix is assumed.
6) The first value of matrix is 1, i.e., A(1,1) = 1.
7) The nth entry of matrix is also 1, i.e., A(n,n) = 1.
8) The first entry of the column vector is 1, i.e., b(1) = 1.
9) The final entry of the column vector is 1, i.e., b(n) = 1.
10) The value of h is assumed to be changed. For this h is treated as a difference of second entry minus first entry, third entry minus second entry and so on.
11) The remaining entries of the matrix are defined in such a way that the loop starts from i = 2 to n – 1, because first and last entries are defined above. The lower diagonal matrix is written in the form:

$A(i,i - 1) = h(x,i - 1)$ (according to the given formula), for i = 2,...n – 1 we will get the lower diagonal entries.

12) The diagonal entries are represented by

$A(i,i) = 2 * (h(x,i - 1) + h(x,i))$ (according to the given formula).

For i = 2,...n – 1 we will get the diagonal entries.

13) The upper diagonal entries are represented by

$A(i,i + 1) = h(x,i)$ (according to the given formula),

for i = 2,...n – 1 we will get the upper diagonal entries.

14) The values of column vector b are

$b(i) = ((3/h(x,i)) * (f(i + 1) - f(i)) - (3/h(x,i - 1)) * (f(i) - f(i - 1)))$ (according to the given formula),

for i = 2,...n – 1 we will get the required entries.

15) The unknown C's are calculated by:
c = A\b.
16) The remaining unknown B's and D's are calculated by writing them in the following form:

$bb(i) = ((f(i + 1)-f(i))/h(x,i))-(h(x,i)/3) * (2 * c(i) + c(i + 1))$,
$dd(i) = (c(i + 1)-c(i))/(3 * h(x,i))$.
1. The loop starts from 1 and ends at n – 1.

17) Finally, the main cubic formula is implemented which will pick the values of C's, B's, D's and f's:

$ss(i) = f(i) + bb(i) * (xx - x(i)) + c(i) * ((xx - x(i)) \wedge 2) + dd(i) * ((xx - x(i)) \wedge 3)$ for i = 1 to n – 1.

```
% Mfile
function [ss]= cubic_spline(b,x,f,xx)
% Inputs and outputs are defined.
% x and f are given data
% xx is the value which is to be calculated
n=length(b);
b=b'; % b is a column vector
f=f'; % f is a column vector
A=zeros(n,n);  % n*n is a matrix
A(1,1)=1; % First entry of matrix is 1
A(n,n)=1; % Last entry of matrix is 1
b(1)=0;  % First entry of b is zero (Natural Spline)
b(n)=0;  % Last entry of b is zero (Natural Spline)
   function h=h(x,i) % Entries of matrix depends on h
     h=x(i+1)-x(i);
end
```

```
for i=2:n-1 % First entry and last entry will be 1 and the remaining
% entries are
  A(i,i-1)=h(x,i-1); % Lower diagonal entries
  A(i,i)=2*(h(x,i-1)+h(x,i)); % Diagonal entries
  A(i,i+1)=h(x,i); % Upper diagonal entries
b(i)=((3/h(x,i))*(f(i+1)-f(i))-(3/h(x,i-1))*(f(i)-f(i-1))); % Column
% vector b
end
c=A\b; % Eliminating ' C's
for i=1:n-1
  % Calculating ' B's and ' D's
bb(i)=((f(i+1)-f(i))/h(x,i))-(h(x,i)/3)*(2*c(i)+c(i+1));
  dd(i)=(c(i+1)-c(i))/(3*h(x,i));
  % Finally, all values are calculated from polynomial
ss(i)=f(i)+bb(i)*(xx-x(i))+c(i)*((xx-x(i))^2)+dd(i)*((xx-x(i))^3);
end
end
```

```
> % Command Window
>> cubic_spline([0 0 0 0],[2.5 4 6.5 8.5],[2 0.5 2 0.1],5)
ans =
   1.3530  0.9567  -0.1761
```

PROBLEM SET 3

1. The values of $sin(x)$ for different values of x are mentioned in degrees:

x (degrees)	20	25	30	35	40
sin(x)	0.342	0.422	0.500	0.573	0.643

Find the value of $sin\left(34^0\right)$ by using forward difference.

2. The table below shows the values of $y = tan(x)$ for $0.1 \le x \le 0.3$.

x	0.1	0.15	0.20	0.25	0.30
y	0.1	0.15	0.20	0.255	0.309

Find the value of $tan(0.27)$ by using forward difference.

3. From below table calculate $f(0.23)$ and $f(0.29)$.

x	0.20	0.22	0.24	0.26	0.28	0.30
f(x)	1.659	1.669	1.680	1.691	1.702	1.713

4. The density of air changes with height in the following manner:

h (km)	0	3	6
ρ (kg/m^3)	1.225	0.905	0.652

Calculate the height at 5.9.

5. From the given table, use divided difference method and calculate Newton's interpolating polynomial by using MATLAB.

x	1	2	3	5
y	3	5	9	11

6. In MATLAB, use divided difference method to calculate the polynomial of degree ≤ 5, that interpolates the table.

x	1	2	3	4	5	6
f(x)	14.6	19.6	30.6	53.6	94.6	159.6

By using that polynomial calculate the value at f(3.4).

7. We have given a function $f(x) = 2x^3 + 4x^2 + 2x + 1$. By using divided difference method find a polynomial that interpolates the function at $x = -2, -1, 0, 1, 2, 3, 4$.
8. By using the points $x_0 = 1, x_1 = 2$ and $x_2 = 3$, find the Lagrange interpolating polynomial for $f(x) = 1/x$ by using MATLAB.
9. By using MATLAB, construct a code for Lagrange interpolating polynomials for the functions:
   (i) $f(x) = cos(4x)e^{3x}$, where $x_0 = 0, x_1 = 0.2$ and $x_2 = 0.6$.
   (ii) $f(x) = ln(x)$, where $x_0 = 2, x_1 = 2.3$ and $x_2 = 3$.
   (iii) $f(x) = cos(x) + 2$, where $x_0 = 0, x_1 = 0.2$ and $x_2 = 0.6$.
10. With the help of MATLAB, construct a code of Neville's method for the following data:

x	8.1	8.3	8.6	8.8
f(x)	16.93	17.65	18.42	18.76

and calculate f(8.7).

11. Develop a code of Neville's method for the following functions and values:
    (i) $f(x) = 2^x$ and the values are $x_0 = 0, x_1 = 2, x_2 = 7$ and $x_3 = 9$.
    (ii) $f(x) = \sqrt{x}$ and the values are $x_0 = 0, x_1 = 1, x_2 = 3$ and $x_3 = 7$.
    (iii) $f(x) = sin(2x)$ and the values are $x_0 = 0, x_1 = 2, x_2 = 4$ and $x_3 = 6$.
12. Find the natural cubic spline of the following data:

x	0	1	2
y	0	2	1.1

13. From the given data:

x	1.1	2.1	3.1	4.1	5.1
y	13.1	15.1	12.1	9.1	13.1

Calculate the natural cubic spline at x =2 .4.

14. The drag coefficient μ_0 of a sphere is represented as a function of Reynolds number Re as mentioned in the table. Calculate the value of drag coefficient μ_0 at Re = 10.

Re	2	200	2000	20000
μ_0	13.90	0.80	0.40	0.43

4 Root Finding Methods

For a given function $f(x)$, the value of x in which $f(x) = 0$ is known as root or zero of the equation. The roots of quadratic equations can be solved analytically but for many higher order equations, it is very difficult or impossible to calculate the zeros. For example, assume a solid sphere which floats in water, according to Archimedes' principle the buoyancy force is equal to the weight of the replaced liquid. Suppose that the volume of the sphere is $V_s = \left(\dfrac{4}{3}\right)\pi r^3$, let V_m be the volume of the displaced water when the sphere is dipped (as shown in Figure 4.1). In static equilibrium, the buoyancy force and the weight of the sphere are balanced, i.e.,

$$\rho_s g V_s = \rho_m g V_m. \tag{4.1}$$

Where g is acceleration due to gravity, ρ_s is the density of the sphere material and ρ_m is the density of the water. The displaced water's volume V_d when the sphere is dipped to a depth d is

$$V_d = \frac{\pi}{3} d^2 (3r - d). \tag{4.2}$$

By Archimedes' principle, the equation in terms of d is

$$d^3 - 3rd^2 + 4\rho r^3 = 0. \tag{4.3}$$

For the given values of specific gravity of the sphere material $\rho = \dfrac{\rho_s}{\rho_m}$ and radius r, the solutions of d can be calculated by one of the root finding methods studied in this chapter. The root finding methods include the bracketing and open methods. In bracketing methods, bisection and *regula falsi* methods are studied. Whereas, in open methods Newton Raphson and secant methods are given full consideration.

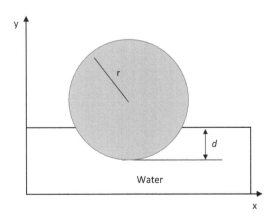

FIGURE 4.1   Dipped sphere.

DOI: 10.1201/9781003385288-4

4.1 BISECTION METHOD

This method is based on the application of the intermediate value theorem. In this method we subdivide an interval in which we know the function has a zero. Suppose an interval s = [a,b] and assume that this interval contains the root, since the sign of the function changes between a and b. Suppose $a_1 = a$ and $b_1 = b$ and assume that c be the midpoint of [a,b], then

$$c_1 = \frac{a_1 + b_1}{2}. \tag{4.4}$$

The approximate solution lies either in (a_1, c_1) or (c_1, b_1). From these two subintervals, one subinterval is chosen. If

$$f(a_1)f(c_1) < 0, \tag{4.5}$$

then the root lies between (a_1, c_1). Or if the sign is opposite then the root lies between (c_1, b_1)

$$f(c_1)f(b_1) < 0. \tag{4.6}$$

Again the interval is divided into two subintervals and the same process is repeated until the value is repeated. The graphical illustration of this procedure is shown in Figure 4.2.

Initially the interval [a,b] has length b−a. After calculating the first approximation the length of the interval reduces to $\frac{b-a}{2}$ and in a similar manner it reduces to $\frac{b-a}{4}$ and so on. The procedure is repeated until the desired small interval, say ϵ, which contains the root is achieved. Thus, the nth interval has length $\frac{|b-a|}{2^n}$. We have

$$\frac{|b-a|}{2^n} \leq \epsilon,$$

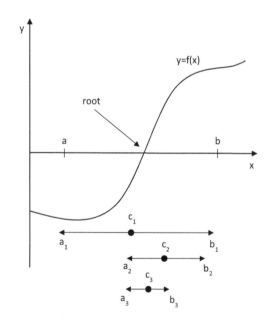

FIGURE 4.2 A graphical behavior of the bisection method.

after simplification

$$n \geq \frac{\log_e \left(|b - a| / \in \right)}{\log_e (2)}.$$ (4.7)

Eq. (4.7) gives the number of iterations needed to achieve an accuracy.

Example 4.1: By using the bisection method calculate the root of $f(x) = x^3 - 5x + 1$ that lies between (0,1).

Solution: Since a = 0 and b = 1, so

$$f(0) = 1, f(1) = -3,$$

hence $c = \dfrac{a+b}{2} = \dfrac{0+1}{2} = 0.5$. So $f(c) = f(0.5) = -1.3750$.

By the statement of bisection method, $f(0.5)f(0) < 0$, so the root lies between (0,0.5).

Again, calculate c, which is $c = \dfrac{0+0.5}{2} = 0.25$. And $f(c = 0.25) = -0.2344$, which shows that $f(0)f(0.25) < 0$, so the root lies between (0,0.25).

Again, calculate c, which is $c = \dfrac{0+0.25}{2} = 0.125$.

And $f(c = 0.125) = 0.3770$, which shows that $f(0.25)f(0.125) < 0$, so the root lies between (0.125,0.25). Again, calculate c, which is $c = \dfrac{0.125 + 0.25}{2} = 0.1875$.

And $f(c = 0.1875) = 0.0691$, which shows that $f(0.1875)f(0.25) < 0$, so the root lies between (0.1875,0.25).

Again, calculate c, which is $c = \dfrac{0.1875 + 0.25}{2} = 0.21875$.

$f(c = 0.21875) = -0.0833$, which shows that $f(0.1875)f(0.21875) < 0$, so the root lies between (0.25,0.21875). Calculate c which is $c = \dfrac{0.21875 + 0.1875}{2} = 0.2031$.

So the approximate root is 0.2031.

Main steps for solving the bisection method in MATLAB: Two scripts are used in the bisection method. One is used to create a general code of bisection (for a single equation) and the second script is used to write the equation in the form of a function.

1) For general code (first script), the inputs and outputs are defined. The required root x and number of iterations are outputs. In inputs, 'func' is a function that is going to evaluate by calling command (@), a and b are the given values between which the root lies and n is the number of iterations that we want to perform.
2) The loop is implemented for refined values.
3) In the bisection method, the functions are defined for a and b, i.e.,
   fa = func(a),
   fb = func(b).
   Here the function defined in input 'func' is evaluated from the second script by using calling command in the command window.
4) In the next step, c is defined, i.e., c = (a + b)/2.
5) This value of c is substituted in the given function to calculate f(c), i.e.,
   fc = func(c).

6) Next we have to choose by either taking fa and fc or fb and fc, in which the product of any two functions is less than zero. For this, 'if' and 'elseif' command is implemented. That is, if the product of fa.fc < 0 then b is replaced by c (b = c) and function of b is replaced by c. This condition runs in such a way that if the statement is true then the statement will be blocked, otherwise it will break the block and move forward towards the next statement.

7) At the end x = c is the required root.

8) Now the second script is opened, the input and output are defined. In this code, x is treated as input and y is treated as output. The given function is written whose root is to be evaluated.

9) Now the command window is used to run both scripts at the same time. The first script inputs and outputs are implemented in the command window. The command @ is used to call the second script function in the command window. In this way, both scripts are used.

```
% Mfile 1
function [x,numiter]=Bisection(func,a,b,n)
% Required root x and number of iterations are outputs
% func is a given function, a and b are the values between
% which the root lies and n is the total number of
% iterations
for numiter=1:n
fa=func(a); % Value of a is used in function
fb=func(b); % Value of b is used in function
c=(a+b)/2; % Average of a and b is c
fc=func(c); % Average value is substituted in function
if fa*fc<0 % Product of two functions less than zero
    b=c;  % Replace b by c
     fb=fc; % Replace function value b by c
elseif fb*fc<0 % If above statement is false then the
   % second statement is used
    a=c;  % Replace a by c
     fa=fc; % Replace function value a by c
end
x=c;  % Required root
end
```

```
% Mfile 2
function y=abc(x)
y=x^3-5*x+1;
end
```

```
>> % Command Window
>> [x,numiter]=Bisection(@abc,0,1,6)
x =
    0.2031
numiter =
         6
```

4.2 *REGULA FALSI* METHOD

Consider an equation of the form $f(x) = 0$ and let $f(a_1)$ and $f(b_1)$ have opposite signs. Connect the points $(a_1, f(a_1))$ and $(b_1, f(b_1))$ by a straight line (as shown in Figure 4.3). The equation of the chord

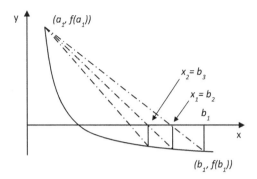

FIGURE 4.3 A graphical behavior of the false position.

joining these two points is $\dfrac{y-f(b_1)}{x-b_1}=\dfrac{f(b_1)-f(a_1)}{b_1-a_1}$. The intersection of the chord with the x-axis gives an approximate value of the root $f(x)=0$. Putting $y=0$ in the chord equation and solving gives

$$x=\frac{a_1 f(b_1)-b_1 f(a_1)}{f(b_1)-f(a_1)}. \tag{4.8}$$

This is the approximate value of the root, assume $x=x_1$. The root lies between $f(x_1)$ and $f(a_1)$ or lies between $f(x_1)$ and $f(b_1)$.

If

$$f(a_1)f(x_1)<0, \tag{4.9}$$

then the root lies between (a_1,x_1). Or if the sign is opposite then the root lies between (x_1,b_1)

$$f(x_1)f(b_1)<0. \tag{4.10}$$

Again, the interval is divided into two subintervals and the same process is repeated until the required root is achieved. The stopping criteria of *regula falsi* is $|x_{n+1}-x_n|<\epsilon$, where ϵ is tolerance.

Example 4.2: By using the *regula falsi* method, calculate the root of $x^3-5x+1=0$ that lies between (0,1).

Solution: Let

$$f(x)=x^3-5x+1,$$

since $f(0)=1, f(1)=-3$. So $x_1=\dfrac{0(-3)-1(1)}{-3-1}=0.2500$, implies that $f(x_1)=f(0.2500)=-0.2344$.

Since $f(0)$ and $f(0.2500)$ have opposite signs. So $f(0)\,f(0.2500)<0$.

So, the root lies between $(0,0.2500)$. Hence $x_2=\dfrac{0(-0.2344)-0.2500(1)}{-0.2344-1}=0.2025$, implies $f(x_2)=f(0.2025)=-0.0044$.

Since $f(0)$ and $f(0.2025)$ have opposite signs.

So, the root lies between (0,0.2025). Since $f(0)$ $f(0.2025) < 0$, hence

$$x_3 = \frac{0(-0.0044) - 0.2025(1)}{-0.0044 - 1} = 0.2017.$$

So for two decimal places the root is 0.20.

Main steps for solving the *regula falsi* **method in MATLAB**: For *regula falsi*, the same procedure is implemented as used for the bisection method. The only difference is that in the bisection method c is calculated by average of a and b, but in this method c is replaced by $x = \dfrac{af(b) - bf(a)}{f(b) - f(a)}$.

For this method two scripts are used. One is used to create a general code of *regula falsi* (for a single equation) and the second script is used to write the equation in the form of a function.

1) For general code (first script), the inputs and outputs are defined. The required root x and number of iterations are outputs. In inputs, 'func' is a function that we are going to evaluate by calling command (@), a and b are given values between which the root lies and n is the number of iterations that we want to perform.
2) A loop is implemented for modification in each iteration.
3) The functions depending on *a* and *b* are defined, i.e.,

fa = func(a),

fb = func(b).

4) In the next step, c is defined, i.e.,

c = (a * fb – b * fa)/(fb – fa).

5) The value of c is used in the function, i.e.,

fc = func(c).

6) Next, we have to choose by either taking fa and fc or fb and fc, in which product of any two functions is less than zero. For this, 'if' and 'elseif' command is implemented. That is, if the product of fa.fc < 0 then b is replaced by c (b = c) along with fb = fc. This condition runs in such a way that if the statement is true then the statement will be blocked, otherwise it will break the block and move forward towards the next statement, which is 'elseif' condition in which fb.fc < 0 replace a by c (a = c) along with fa = fc.
7) At the end x = c is required root.
8) Now the second script is opened, the input and output are defined. In this code, x is treated as input and y is treated as output. The given function is written whose root is to be evaluated. Save the script by some name.
9) Now the command window is used to run both scripts at the same time. The first script inputs and outputs are implemented in the command window.

```
% Mfile 1
function [x,numiter]=False_Position(func,a,b,n)
% Required root x and number of iterations are outputs
% func is a given function, a and b are the values between
% which the root lies and n is the total number of
%iterations
for numiter=1:n
fa=func(a); % Value of a is used in function
fb=func(b); % Value of b is used in function
c=(a*fb-b*fa)/(fb-fa); % Only difference between
%the bisection method and the regula falsi is c
fc=func(c); % Average value is substituted in the function
if fa*fc<0 % Product of two functions less than zero
    b=c;  % Replace b by c
    fb=fc; % Replace function value b by c
```

```
    elseif fb*fc<0 % If the above statement is false than the
    % second statement is used
        a=c;   % Replace a by c
        fa=fc; % Replace function value a by c
    end
    x=c;   % Required root
end
```

```
% Mfile 2
function y=abc(x)
y=x^3-5*x+1;
end
```

```
>> % Command Window
>> [x,numiter]=False_Position(@abc,0,1,4)
x =
      0.2016
numiter =
            4
```

4.3   NEWTON RAPHSON METHOD

This method is the best root finding method. The reason is that it is simple and fast. For the derivation of the Newton Raphson method, consider a point x_1 on the x-axis (as shown in Figure 4.4). The location of point x_1 on the curve is $(x_1, f(x_1))$. From this point on the curve draw a tangent line and assume that it intercepts x-axis at x_2. For this the equation of the tangent line is

$$y - f(x_1) = f'(x_1)(x - x_1). \tag{4.11}$$

As the tangent line intercepts the x-axis, substituting $y = 0$ and $x = x_2$ in Eq. (4.11) yields

$$x_2 = x_1 - \frac{f(x_1)}{f'(x_1)}, \tag{4.12}$$

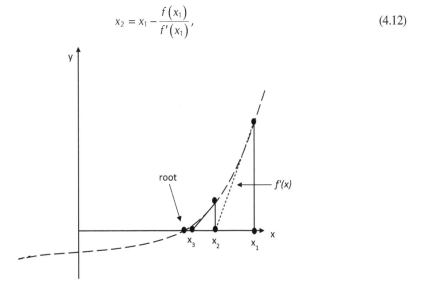

FIGURE 4.4   A graphical behavior of the Newton Raphson method.

as x_2 is calculated the location of this point on the curve is $(x_2, f(x_2))$. Again, draw a tangent line at that point on the curve and assume that it intercepts the x-axis. We get

$$x_3 = x_2 - \frac{f(x_2)}{f'(x_2)},$$
(4.13)

this process is repeated until the required root is obtained. The general form is

$$x_{n+1} = x_n - \frac{f(x_n)}{f'(x_n)}.$$
(4.14)

The stopping criteria of the Newton Raphson method is the same as calculated for the *regula falsi* method.

The disadvantage of this method is that the derivative of a given function is sometimes difficult to calculate (as shown in Figure 4.5). Secondly, the tangent line is not always a satisfactory approximation of the function (as shown in Figure 4.6).

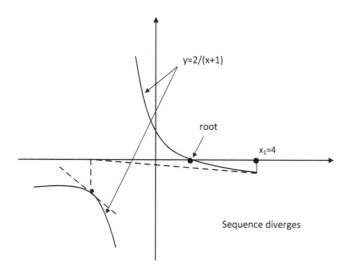

FIGURE 4.5   A graphical behavior of the divergence sequence.

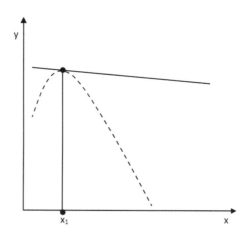

FIGURE 4.6   A graphical behavior of the divergence sequence.

Definition 4.1: Let us assume a sequence $x_1 x_2 \ldots$ that converges to a number 'a' and set $\in_n = a - x_n$. If there exists a positive constant s and a number q such that

$$\lim_{n \to \infty} \frac{|\in_{n+1}|}{|\in_n|^q} = s. \tag{4.15}$$

In Eq. (4.15), s is the asymptotic error constant and q is the order of convergence of the sequence. A root of order 1 is known as a simple root.

Now moving forward towards the convergence criteria of the Newton Raphson method for simple roots. From Eq. (4.14) we have

$$\in_{n+1} = a - x_{n+1} = a - x_n + \frac{f(x_n)}{f'(x_n)}. \tag{4.16}$$

If $f''(x)$ exists, then by expanding f about x_n (using Taylor series)

$$f(a) = f(x_n) + (a - x_n)f'(x_n) + \frac{(a - x_n)^2}{2}f''(\Pi_n). \tag{4.17}$$

Where Π_n lies between x_n and a. As $f(a) = 0$ and $f'(a) \neq 0$, substituting the value of $f(x_n)$ from Eq. (4.16) into (4.17), we get

$$\in_{n+1} = -\in_n^2 \frac{f''(\Pi_n)}{2f'(x_n)}. \tag{4.18}$$

Eq. (4.18) shows that if a is a simple root and $f'(a) \neq 0$ then Newton Raphson converges quadratically.

Example 4.3: By using the Newton Raphson method, calculate the root of $x^3 - 5x + 1 = 0$.

Solution: By using formula,

$$x_{i+1} = x_i - \frac{x_i^3 - 5x_i + 1}{3x_i^2 - 5}.$$

Assume that $x_1 = 0.1$.

Calculate the values for i = 1, 2, 3 ... until the value is repeated.

For i = 1,

$$x_2 = x_1 - \frac{x_1^3 - 5x_1 + 1}{3x_1^2 - 5} = 0.1 - \frac{(0.1)^3 - 5(0.1) + 1}{3(0.1)^2 - 5} = 0.2008.$$

For i = 2,

$$x_3 = x_2 - \frac{x_2^3 - 5x_2 + 1}{3x_2^2 - 5} = 0.2008 - \frac{(0.2008)^3 - 5(0.2008) + 1}{3(0.2008) - 5} = 0.2016.$$

For i = 3,

$$x_4 = x_3 - \frac{x_3^3 - 5x_3 + 1}{3x_3^2 - 5} = 0.2016 - \frac{(0.2016)^3 - 5(0.2016) + 1}{3(0.2016)^2 - 5} = 0.2016.$$

So, the root is 0.2016.

Main for solving the Newton Raphson method in MATLAB: For the Newton Raphson method three scripts are used. The first script is used for general Newton's Raphson code, the second one is used for a given function and the third one is used for the derivative of the function. The main steps are:

1) For the first script, outputs and inputs are defined: c and 'numiter' are outputs and func, dfunc, x and n are inputs. The 'numiter' symbol stands for number of iterations, func is the function, dfunc is used for derivative of the function, and x is initial value which is to be assigned and n stands for number of iterations.
2) A loop is implemented from 1 to n.
3) The function and the derivatives are defined, i.e.,
   f = func(x),
   df = dfunc(x).
   These are defined in a loop and one more thing is added in the loop and that is x, i.e.,
   x = x − (f/df).
4) What will happen by implementing a loop? It starts from 1, the first given value of x is delivered to the f = func(x) and then moves towards df = dfunc(x) and all these values are moved to the formula x = x − (f/df). The first iteration is completed, and the first value of x is calculated. This value is used in the second iteration and the process is repeated till the selected value of n. At last, c = x is written as output.
5) The second script is opened, and a given function is written and saved in the form of any name.
6) The third script is opened, and the derivative of the function is written and saved in the form of any name.
7) Now we move to the command window, the first script is used as the main tool, the outputs and inputs are written in the command window. The func is replaced by @ with second script name and dfun is replaced with @ with third script name. The initial guess x and n number of iterations are defined.

```
% Mfile 1
function [c,numiter]=Newton_Raphson(func,dfunc,x,n)
% Required root c and number of iterations are outputs
% func is a given function, a and b are the values
% between which the root lies, and n is the total number of
% iterations
for numiter=1:n
    f=func(x); % Given function
    df=dfunc(x); % Derivative of the function
    x=x-(f/df); % Given formula of Newton Raphson method
end
c=x; % Required root
```

```
% Mfile 2
function y=abc(x)
y=x^3-5*x+1;
end
```

```
%Mfile 3
function y=der_abc(x)
y=3*x^2-5;
end
```

```
>> % Command Window
>> [c,numiter]=Newton_Raphson(@abc,@der_abc,0.1,5)
c =
     0.2016
numiter =
     5
```

4.4 SECANT METHOD

In the Newton Raphson method, the evaluation of the derivative is the main problem. Although for polynomials and some other functions it is not difficult to calculate the derivative, there are some functions whose derivatives are difficult to calculate. That's why the secant method is used. For this consider two points $\left(x_1, \dot{f}(x_1)\right)$ and $\left(x_2, \dot{f}(x_2)\right)$ on the curve (as shown in Figure 4.7) and draw a secant line that connects them. The secant line intercepts the x-axis at x_3. Next use x_2 and x_3 on the curve and draw a secant line which intercepts the x-axis at x_4. The procedure is repeated until the root is obtained. Mathematically, the derivative of the Newton Raphson method is approximated by a backward finite divided difference, which is:

$$\dot{f}'(x_i) \cong \frac{\dot{f}(x_i) - \dot{f}(x_{i-1})}{x_i - x_{i-1}}. \tag{4.19}$$

Substituting this expression in the Newton Raphson formula, we get

$$x_{i+1} = x_i - \frac{\left(x_i - x_{i-1}\right)\dot{f}(x_i)}{\dot{f}(x_i) - \dot{f}(x_{i-1})}. \tag{4.20}$$

This formula is known as the secant method.

Suppose a be a simple root, then the convergence rate of the secant method is [3-83]

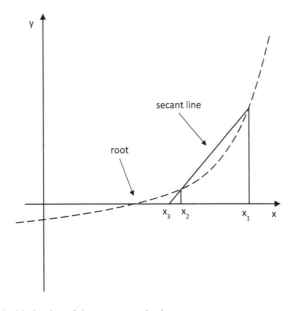

FIGURE 4.7 A graphical behavior of the secant method.

$$\lim_{n \to \infty} \frac{|\epsilon_{n+1}|}{|\epsilon_n|^{1.618}} = \left| \frac{\dot{f}''(a)}{2\dot{f}'(a)} \right|^{0.618} \neq 0. \tag{4.21}$$

For the simple root, 1.618 is the convergence rate for the secant method. Thus, the sequence converges faster than the linear but less fast than the quadratic.

Example 4.4: By using the secant method, calculate the root of $x^3 - 5x + 1 = 0$ that lies between (0,1).

Solution: By using the formula:

$$x_{i+1} = x_i - \frac{(x_i - x_{i-1})\dot{f}(x_i)}{\dot{f}(x_i) - \dot{f}(x_{i-1})}, \text{for } i = 2, 3, 4 \ldots.$$

Moreover, $x_1 = 0$ and $x_2 = 1$.

For i = 2,

$$x_3 = x_2 - \frac{(x_2 - x_1)\dot{f}(x_2)}{\dot{f}(x_2) - \dot{f}(x_1)} = 1 - \frac{(1-0)(-3)}{-3-1} = 0.25.$$

For i = 3,

$$x_4 = x_3 - \frac{(x_3 - x_2)\dot{f}(x_3)}{\dot{f}(x_3) - \dot{f}(x_2)} = 0.25 - \frac{(0.25-1)(-0.234)}{-0.234+3} = 0.186.$$

For i = 4,

$$x_5 = x_4 - \frac{(x_4 - x_3)\dot{f}(x_4)}{\dot{f}(x_4) - \dot{f}(x_3)} = 0.186 - \frac{(0.186-0.25)(0.0764)}{0.0764+0.234} = 0.2018.$$

After performing one more iteration the value is repeated for three decimal places. The root will be x = 0.202.

Main steps for solving the secant method in MATLAB: For the secant method two scripts are used. The first script is used for code and the second one is used for a given function. The main steps are:

1) For the first script, outputs and inputs are defined: c (root) and number of iterations 'numiter' are the outputs and func, x1 and x2 are the inputs. The function is defined by 'func' and x1 and x2 are the initial guesses.
2) As the formula is

$$x_{i+1} = x_i - \frac{(x_i - x_{i-1})\dot{f}(x_i)}{\dot{f}(x_i) - \dot{f}(x_{i-1})}, \text{for } i = 2, 3, 4 \ldots.$$

In order to calculate the root of the function, the loop is implemented and, in that loop, x1 and x2 are replaced by some other variables (s and r). This step is performed because x1 and x2 are refined in each iteration.
3) In a similar manner, the functions (func) are defined in the form of q and u. Note that these functions are also refined in each iteration.
4) Next the secant formula is used (represented by u3).
5) After this the first iteration is not completed yet. The values are refined in this iteration by replacing x1 by r (the second value will be the first one) and x2 by u3 (the calculated

value will be the second one). The loop is ended after this process. The procedure is repeated until the value is repeated.

6) At the end x is replaced with c (output).

7) The second script is used to write the given function and is saved by any name.

8) At the end, the command window is used to run both scripts.

```
% Mfile 1
function [c,numiter]=Secant(func,x1,x2)
% Required root c and number of iterations are outputs
% func is the given function, x1 and x2 are the values
% between which the root lies and n is the total number of iterations
for numiter=1:6
    s=x1;
    r=x2;
    q=func(s);
    u=func(r);
    u3=r -((r-s)/(u-q))*(u); % Secant formula
    x1=r; % Refined value
    x2=u3; % Refined value
end
c=u3;
end
```

```
% Mfile 2
function y=abc(x)
y=x^3-5*x+1;
end
```

```
>> % Command Window
>> [c,numiter]=Secant(@abc,0,1)
c =
    0.2016
numiter =
    6
```

4.5   NEWTON RAPHSON METHOD FOR THE SYSTEM OF EQUATIONS

In order to solve the Newton Raphson method for the system of equations, the Taylor series is expanded about the point x:

$$f_i(x+\Delta x) = f_i(x) + \sum_{j=1}^{n} \frac{\partial f_i}{\partial x_j} \Delta x_j + O(\Delta x)^2. \tag{4.22}$$

After dropping the terms of $O(\Delta x)^2$, the Eq. (4.22) becomes:

$$f(x+\Delta x) = f(x) + J(x)\Delta x. \tag{4.23}$$

Where $J(x)$ is the Jacobian for the $n\times n$ matrix which is

$$J(x_1, x_2, x_3 \ldots x_n) = \begin{bmatrix} \dfrac{\partial f_1}{\partial x_1} & \cdots & \dfrac{\partial f_1}{\partial x_n} \\ \cdot & \cdots & \cdot \\ \dfrac{\partial f_n}{\partial x_1} & \cdots & \dfrac{\partial f_n}{\partial x_n} \end{bmatrix}. \tag{4.24}$$

Set $f(x + \Delta x) = 0$, we get

$$J(x)\Delta x = -f(x),$$

or

$$\Delta x = J^{-1}(x) f(x). \tag{4.25}$$

Next, we will calculate the iteration by

$$x_1 = x_0 + \Delta x = x_0 + J^{-1}(x) f(x). \tag{4.26}$$

This process is repeated until the required root is achieved.

Example 4.5: Solve the nonlinear system

$$x^2 + 4y^2 - 4 = 0, x^2 - 2x - y + 0.5 = 0.$$

By using the Newton Raphson method with a starting point (0.25,2).

Solution: The vector function and Jacobian matrix are

$$f(x,y) = \begin{bmatrix} x^2 + 4y^2 - 4 \\ x^2 - 2x - y + 0.5 \end{bmatrix}, J(x,y) = \begin{bmatrix} 2x & 8y \\ 2x - 2 & -1 \end{bmatrix}.$$

The differentials Δx and Δy are solutions of the linear system

$$\begin{bmatrix} 0.5001 & 16.0004 \\ -1.4999 & -1 \end{bmatrix} \begin{bmatrix} \Delta x \\ \Delta y \end{bmatrix} = -\begin{bmatrix} 12.0625 \\ -1.9375 \end{bmatrix},$$

or

$$\begin{bmatrix} \Delta x \\ \Delta y \end{bmatrix} = -\begin{bmatrix} 0.5001 & 16.0004 \\ -1.4999 & -1 \end{bmatrix}^{-1} \begin{bmatrix} 12.0625 \\ -1.9375 \end{bmatrix}.$$

Simplifying gives

$$\Delta X = \begin{bmatrix} \Delta x \\ \Delta y \end{bmatrix} = \begin{bmatrix} -0.8059 \\ -0.7287 \end{bmatrix}.$$

The next point is

$$X_1 = X_0 + \Delta X = \begin{bmatrix} -0.5559 \\ 1.2713 \end{bmatrix}.$$

Similarly, the next four iterations are

$$X_2 = \begin{bmatrix} 0.0044 \\ 1.0804 \end{bmatrix}, X_3 = \begin{bmatrix} -0.3683 \\ 1.0502 \end{bmatrix}, X_4 = \begin{bmatrix} -0.1257 \\ 1.0084 \end{bmatrix}, X_5 = \begin{bmatrix} -0.1257 \\ 1.0084 \end{bmatrix}.$$

Main steps for solving the Newton Raphson method for the system of equations in MATLAB:
The Newton Raphson method for the system of equations consists of two scripts. The first script stands for general code and the second one is used for given functions. The main steps are:

1) For the first script, outputs and inputs are defined: roots and 'numiter' are outputs and the function 'func', vector v (given initial values) and x (given initial values) are inputs.
2) As

$$\Delta x = J^{-1}(x) f(x).$$

   To get this relation, we have to calculate the inverse of the Jacobian and as we know that the Jacobian is a matrix, we need initial values as a column vector. So first of all x = x' is taken in the next step.
3) For roots, loop is implemented.
4) Next the function 'func' is defined, i.e.,
   f0 = func(x).
5) Secondly, the Jacobian is calculated by using the build in command of the Jacobian.
6) After that, dx is calculated, i.e.,
   dx = inv(jac) * (–f0).
7) Next dx is added with the given initial guess x, i.e.,
   x = x + dx.
8) At the end, the vector x is replaced with the output 'roots' and the loop is ended.
9) This step is repeated again and again until the required root is achieved.
   Note: In the current problem the 'numiter' is taken to be 6, in order to get better visualization of the roots the 'numiter' can be increased.
10) In the second script, the given system is written if the functions depend on x and y then replace x by x(1) and y by x(2).
11) After saving this file, the command window is used to run both files.

```
% Mfile 1
function [roots,numiter]=Newton_Raphson_System(func,x,v)
% Outputs and inputs are defined: roots and 'numiter' are
% the outputs and the function 'func' which is saved in
% the second script plus vector v and x are inputs
% Given initial values are inputs
x=x'; % The initial vector is a column vector
for numiter= 1:4
    f0 = func(x); % Function value
    jac=jacobian(func,v); % Jacobian is calculated
dx=inv(jac)*(-f0); % From the given relation dx is
% calculated
    x=x+dx;
    roots=x;
end
end
```

```
% Mfile 2
function y=fx2(x)
y=[(x(1)^2)+4*(x(2)^2)-4; (x(1)^2)-2*x(1)-x(2)+0.5];
end
```

```
>> % Command Window
>> [roots,numiter] =Newton_Raphson_System(@fx2,[0.25,2],[0.25,2])
roots =
    -0.1257
     1.0084
numiter =
     4
```

PROBLEM SET 4

1. Use MATLAB to find the root of $y = -1 - x + x^3$ by bisection method.
2. Use single script to calculate the root of $2x - \sqrt{1 + cos(x)} = 0$ by bisection method.
3. By using $a = 0$ and $b = 1$ calculate the root of the following:

$$x^3 + 2x^2 - 3x - 3.$$

4. Use two Mfiles and calculate the root of $e^{-x} - x^3 - x = 0$ by using *regula falsi*.
5. Use *regula falsi* to approximate the solution $e^{-x} sin(x) + x + 5 = 0$ in the interval [2.9,3.9].
6. Use single script to calculate the root of $e^{-x} - x = 0$ by using *regula falsi*.
7. Test your program by using the Newton Raphson method and calculate the roots of the following functions:
   (i) $4 - x - 3x^3 + x^5 = 0$
   (ii) $xsin(x) + e^x = 0$
   (iii) $x^3 - x^2 + e^x = 0.$
8. In a chemical engineering process, H_2O is heated at a very high temperature which causes a split of O_2 and H_2

$$H_2O \Leftrightarrow \frac{1}{2}O_2 + H_2,$$

assume that only this reaction is involved, the mole fraction x of water that splits apart is represented by

$$C = \frac{x}{1-x}\sqrt{\frac{2t_s}{2+x}}.$$

Where t_s represents the total pressure of the mixture and C is the reaction's equilibrium constant. Calculate the value of x if $t_s = 2.3$ and $C = 0.01$ by using Mfile.
9. Consider a hollow horizontal cylinder of radius r filled with the liquid of volume V and length related to the depth of the liquid is represented by d, we have

$$V = \left(r^2 cos^{-1}\left(\frac{r-d}{r}\right) - (r-d)(2rd - d^2)^{1/2}\right)L.$$

We have $L = 3m^3, r = 2m$ and $V = 2.4m^3$, calculate d.

10. Use single script to calculate the root of the following function by using the Newton Raphson method:

$$-2 - 4x - 3x^2 + 2x^3 + x^4 = 0.$$

11. Without using the function command use the secant method to calculate the root of the following function:

$$0.1x^4 - 3.8x^3 + 1.2x^2 + 1.43x + 0.21 = 0.$$

12. Test the following function by using the secant method

$$sin(x) - 0.1x = 0.$$

13. Consider two equations which describe the intersection of a parabola with a vertex at origin and a unit circle centered at origin.

$$f(x_1, x_2) = x_1^2 + x_2^2 - 1, g(x_1, x_2) = x_1^2 - x_2,$$

calculate the roots of the following functions by using the Newton Raphson method (see Figure 4.8).

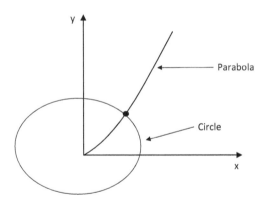

FIGURE 4.8   Intersection of parabola and circle.

14. Find the solution of the following Lorenz equations by using the Newton Raphson code:

$$x_2 - x_1 = 0,$$

$$5x_1 - x_2 - x_1 x_3 = 0,$$

$$x_1 x_2 - 16x_3 = 0.$$

5 Numerical Integration

In this chapter, the problems related to the definite integrals are discussed, i.e.,

$$T = \int_a^b f(x)\,dx. \tag{5.1}$$

Eq. (5.1) shows the integration of a function $f(x)$ with respect to an independent variable x bounded by a and b. Consider a sphere of mass m attached to a spring of free length h (as shown in Figure 5.1). The stiffness of the spring is ϖ, the coefficient of friction between the horizontal rod and the mass is τ. If the sphere is released from $x = h$, then its speed at $x = 0$ is given by

$$S = \sqrt{2\int_0^h \left((\tau g) + \frac{\varpi}{m}(\tau h + x)\left(1 - \frac{h}{\sqrt{h^2 + x^2}} \right) \right) dx}. \tag{5.2}$$

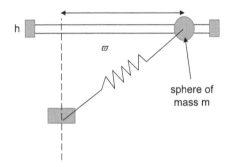

FIGURE 5.1   A graphical behavior of sphere attached to a spring.

Eq. (5.2) can be solved by integrating but there are some functions which cannot be solved analytically, so numerical integration is used.

In this book, three numerical integrating techniques are studied, namely, Newton-Cotes formulas, adaptive methods and Gaussian quadrature.

5.1 NEWTON-COTES FORMULAS

In Newton-Cotes formulas, the function or numerical data is replaced with a polynomial:

$$T = \int_a^b f(x)\,dx \cong \int_a^b P_n(x)\,dx. \tag{5.3}$$

Where $P_n(x)$ is a polynomial of nth order, i.e.,

$$P_n(x) = a_0 + a_1 x^1 + \ldots + a_n x^n. \tag{5.4}$$

The a's are calculated by using the interpolating polynomial which gives the same values of the given function at a finite number of points. Further, the Newton-Cotes formulas are studied for open and closed forms. In open form the upper and lower limits of integration are extended outside the range of data (as shown in Figure 5.2 a). In closed form, the upper and lower limits of integration are known (as shown in Figure 5.2 b). Some of the known Newton-Cotes formulas are

DOI: 10.1201/9781003385288-5   **107** trapezoidal (linear polynomial approximation), Simpson's 1/3 (parabolic polynomial approxima-tion) and Simpson's 3/8 (cubic polynomial approximation).

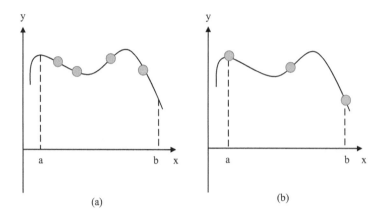

FIGURE 5.2   Open (a) and closed (b) integration.

5.1.1   THE TRAPEZOIDAL RULE

In trapezoidal rule, the function $f(x)$ is approximated by using a first order polynomial which shows that the function is replaced by a straight line in each interval (as shown in Figure 5.3). In this method, the area under each interval is trapezoid. If the areas of the trapezoids are represented by $T_1, T_2 \ldots T_n$, and the step size is denoted by $h = (b-a)/n$, where $n = 1, 2, 3 \ldots$ then

$$T_1 = \left(\frac{f(x_0) + f(x_1)}{2} \right) h, T_2 = \left(\frac{f(x_1) + f(x_2)}{2} \right) h, \ldots T_n = \left(\frac{f(x_{n-1}) + f(x_n)}{2} \right) h.$$

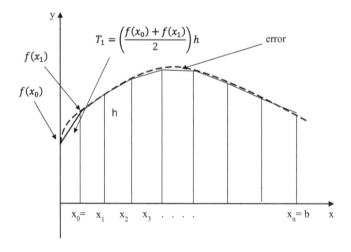

FIGURE 5.3   A graphical illustration of trapezoidal rule.

Therefore, the integral T is approximated by

$$T = \int_a^b f(x) dx \cong \frac{h}{2} \left(f(x_0) + 2f(x_1) + 2f(x_2) + \ldots 2f(x_{n-1}) + f(x_n) \right),$$

or

$$T = \frac{h}{2}\left(\left(f\left(x_0\right)+f\left(x_n\right)\right)+2\left(f\left(x_1\right)+f\left(x_2\right)+\ldots+f\left(x_{n-1}\right)\right)\right). \tag{5.5}$$

Eq. (5.5) is known as the composite trapezoidal rule.

Definition 5.1: An infinite sequence or series involves an error after truncating the series or sequence after a finite number of terms is known as truncation error.

For truncation error, consider a function $y(x)$

$$y(x) = \int_a^x f(x)\,dx. \tag{5.6}$$

By the definition of derivative, we have

$$y'(x) = f(x), y''(x) = f'(x), \ldots. \tag{5.7}$$

Assume that $f(x)$ and $y(x)$ are smooth and continuous in the interval $[x_i, x_{i+1}]$, thus by expanding y_{i+1} about $x = x_i$

$$y_{i+1} = y_i + hy_i' + \frac{h^2}{2!}y_i'' + \frac{h^3}{3!}y_i''' + O\left[h^4\right]. \tag{5.8}$$

Or

$$y_{i+1} - y_i = hy_i' + \frac{h^2}{2!}y_i'' + \frac{h^3}{3!}y_i''' + O\left[h^4\right],$$

$$y_{i+1} - y_i = hf\left(x_i\right) + \frac{h^2}{2!}f'\left(x_i\right) + \frac{h^3}{3!}f''\left(x_i\right) + O\left[h^4\right]. \tag{5.9}$$

Similarly, apply Taylor series about $x = x_i$ yields

$$f\left(x_{i+1}\right) = f\left(x_i\right) + hf'\left(x_i\right) + \frac{h^2}{2!}f''\left(x_i\right) + \frac{h^3}{3!}f'''\left(x_i\right) + O\left[h^4\right],$$

or

$$f'\left(x_i\right) = \frac{f\left(x_{i+1}\right)-f\left(x_i\right)}{h} - \frac{h}{2!}f''\left(x_i\right) - \frac{h^2}{3!}f'''\left(x_i\right) + O\left[h^3\right]. \tag{5.10}$$

Substituting Eq. (5.10) in Eq. (5.9)

$$y_{i+1} - y_i = hf\left(x_i\right) + \frac{h^2}{2!}\left(\frac{f\left(x_{i+1}\right)-f\left(x_i\right)}{h} - \frac{h}{2!}f''\left(x_i\right) - \frac{h^2}{3!}f'''\left(x_i\right)\right) + \frac{h^3}{3!}f''\left(x_i\right) + O\left[h^4\right],$$

or

$$y_{i+1} - y_i = h\frac{f\left(x_{i+1}\right)+f\left(x_i\right)}{2} - \frac{h^3}{12}f''\left(x_i\right) + O\left[h^4\right]. \tag{5.11}$$

In Eq. (5.11), the exact area under the curve is $y_{i+1} - y_i$ and the area of a trapezoid is

$$h\frac{f\left(x_{i+1}\right)+f\left(x_i\right)}{2}.$$

This implies that the truncation error is

$$T_r = -\frac{h^3}{12} f''(x_i) + O[h^4],$$

or more preciously,

$$T_r = -\frac{h^3}{12} f''(\zeta_i), x_i < \zeta_i < x_{i+1}. \tag{5.12}$$

Eq. (5.12) is the truncation error of a single trapezoid. For entire interval, the truncation error is (by the definition of mean $\sum_{i=0}^{n-1} f''(\zeta_i) = nf''_{avg}$)

$$TE = -\frac{h^3}{12} nf''_{avg} = -\frac{h^3}{12}\left(\frac{b-a}{h}\right) f''_{avg} = -\frac{h^2}{12}(b-a) f''_{avg}. \tag{5.13}$$

Eq. (5.13) is the total error.

Example 5.1: Evaluate $T = \int_{-5}^{5} x^{10} dx$ by using the trapezoidal rule.

Solution: Here $y(x) = x^{10}$. Where $b - a = 10$, and $h = \frac{10}{10} = 1$. Hence

x	−5	−4	−3	−2	−1	0	1	2	3	4	5
y	9765625	1048576	59049	1024	1	0	1	1024	59049	1048576	9765625

By formula

$$T = \int_{-5}^{5} x^{10} dx = \frac{1}{2}\left(\begin{array}{l}(9765625 + 9765625) \\ + 2(1048576 + 59049 + 1024 + 1 + 0 + 1 + 1024 + 59049 + 1048576)\end{array}\right),$$

$$T = \int_{-5}^{5} x^{10} dx = 11982925.$$

Main steps for solving the trapezoidal rule in MATLAB: For the trapezoidal rule two scripts are used. One script is used to generate a general code and the second is used for any given function.

1) In the first script, the outputs and inputs are defined. T is the output which is the final value and function 'func', initial value 'a', final value 'b' and the number of intervals 'n' are the inputs.
2) The initial value is defined as x = a.
3) By formula the first and last values are substituted in the given function to get first and last values of the formula, i.e.,
   s = func(a),
   m = func(b).
4) By given formula h is defined as
   h = (b − a)/n.
5) So $\frac{h}{2}\left(\left(y_0 + y_n\right)\right)$ is calculated, now we must calculate $2\left(y_1 + y_2 + \ldots + y_{n-1}\right)$.
6) Before starting the loop t = 0 is implemented.

7) The 'for' loop is implemented, the values of x are incremented by adding h as
   x = x + h.
8) The values $2(y_1, y_2, \ldots y_{n-1})$ are calculated by command:
   r = 2 * func(x).
   The function func(x) will take all the incremented values of x in this command and calculate $2(y_1, y_2, \ldots y_{n-1})$.
   Before ending the loop all these values are added, i.e.,
   t = t + r.
This will add all the values from y_1 to y_{n-1}.
9) At the end the output T is written with the given formula:
   T = (h/2) * (s + m + t).
10) Next the second script is saved by any name and the given function is defined.
11) At the end, the command window is used to run both files.

```
% Mfile 1
function T=trapezoidal(func,a,b,n)
% T is the output
% func is the given function, a is the initial
% value, b is the final value and
% n is the number of intervals
x=a; % Replace first value by x
s=func(a); % First value of the function
m=func(b); % Final value of the function
h=(b-a)/n; % Interval
t=0; % For sequence to be added
for i=1:n-1
  x=x+h; % x in each interval
  r=2*func(x); % Remaining values of the
  %function rather than first and last ones
  t=t+r; % Addition of all values
end
T=(h/2)*(s+m+t); % Final formula
```

```
% Mfile 2
  function y=rst(x)
  y=x^10;
  end
```

```
>> % Command Window
>> T=trapezoidal(@rst,-5,5,10)
T =
 11982925
```

5.1.2 SIMPSON'S 1/3 RULE

In this method, the function in the integral $f(x)$ is replaced by a second order polynomial or a
parabola:

$$S = \int_a^b f(x)\, dx. \tag{5.14}$$

For the determination of a parabola on the curve, three points are needed. Consider the points $(x_{i-1}, f(x_{i-1})), (x_i, f(x_i))$ and $(x_{i+1}, f(x_{i+1}))$ on the curve and assume that the parabola passes through these points (as illustrated in Figure 5.4). The area under the curve is approximated by the area under the parabola. Therefore, the parabola is:

$$P(x) = ax^2 + bx + c. \tag{5.15}$$

The points a, b and c are calculated by

$$f(x_{i-1}) = a(-h)^2 + b(-h) + c, \tag{5.16}$$

$$f(x_i) = c, \tag{5.17}$$

$$f(x_{i+1}) = a(h)^2 + b(h) + c. \tag{5.18}$$

The value of x_i is taken at the origin, i.e., $x = 0$. Solving Eqs. (5.16), (5.17) and (5.18), we will get the values of a, b and c.

$$a = \frac{f(x_{i+1}) - 2f(x_i) + f(x_{i-1})}{2(h)^2}, b = \frac{f(x_{i+1}) - f(x_{i-1})}{2h}, c = f(x_i).$$

Thus, in two segments the area under the curve is

$$S = \left(\frac{f(x_{i+1}) + 4f(x_i) + f(x_{i-1})}{3} \right) h. \tag{5.19}$$

The Eq. (5.19) is known as Simpson's 1/3 formula.

The achieved formula is the integral over the two segments, for the entire domain of equal width h we have

$$S = \left(\frac{h}{3} \right) \left(f(x_0) + 4 \sum_{i=1,3,5...}^{n-1} f(x_i) + 2 \sum_{i=2,4,6...}^{n-2} f(x_i) + f(x_n) \right). \tag{5.20}$$

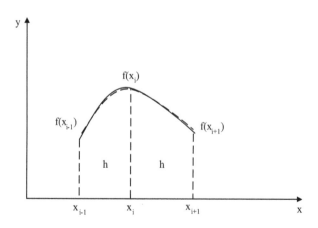

FIGURE 5.4   A graphical illustration of Simpson's 1/3 rule.

The procedure of calculating the truncation error of Simpson's 1/3 formula is similar to the truncation error of the trapezoidal rule. The exact area under the curve between two segments is $y_{i+1} - y_{i-1}$ (as shown in Figure 5.4). By using Taylor series, expanding y_{i+1} and y_{i-1} about the point y_i gives

$$y_{i+1} = y_i + hy_i' + \frac{h^2}{2!} y_i'' + \frac{h^3}{3!} y_i''' + \frac{h^4}{4!} y_i'''' + \frac{h^5}{5!} y_i''''' \quad \dots \tag{5.21}$$

$$y_{i-1} = y_i - hy_i' + \frac{h^2}{2!} y_i'' - \frac{h^3}{3!} y_i''' + \frac{h^4}{4!} y_i'''' - \frac{h^5}{5!} y_i''''' \quad \dots \tag{5.22}$$

Subtracting Eqs. (5.21) and (5.22) and using the Eq. (5.7) yields

$$y_{i+1} - y_{i-1} = 2hf(x_i) + \frac{h^3}{3} f''(x_i) + \frac{h^5}{60} f''''(x_i) + O(h^7). \tag{5.23}$$

By using the finite approximation for $f''(x_i)$ (discussed in Chapter 7)

$$f''(x_i) = \frac{f(x_{i+1}) - 2f(x_i) + f(x_{i-1})}{h^2} - \frac{h^2}{12} f''''(x_i) + O(h^4). \tag{5.24}$$

Putting Eq. (5.24) in Eq. (5.23)

$$y_{i+1} - y_{i-1} = \frac{h}{3} \left(f(x_{i+1}) + 4f(x_i) + f(x_{i-1}) \right) - \frac{1}{90} h^5 f''''(x_i) + O(h^7). \tag{5.25}$$

From Eq. (5.25), the truncation error per step for Simpson's 1/3 rule is

$$T_r = -\frac{1}{90} h^5 f''''(\zeta_i), \text{where } x_{i-1} < \zeta_i < x_{i+1}. \tag{5.26}$$

Eq. (5.26) is the truncation error per step. For entire interval, the truncation error is

$$TE = -\frac{1}{90} h^5 \frac{n}{2} f_{avg}''''(\zeta_i) = -\frac{1}{90} h^5 \frac{b-a}{2h} f_{avg}''''(\zeta_i) = -\frac{1}{180} h^4 (b-a) f_{avg}''''(\zeta_i). \tag{5.27}$$

Eq. (5.27) is the total error.

Example 5.2: Evaluate $S = \int_{-5}^{5} x^2 dx$ by using Simpson's 1/3 rule.

Solution: Here $y(x) = x^2$. Assume h = 1. Hence:

x	−5	−4	−3	−2	−1	0	1	2	3	4	5
y	25	16	9	4	1	0	1	4	9	16	25

By formula

$$S = \int_{-5}^{5} x^2 dx = \frac{1}{3} \left((25 + 25) + 4(16 + 4 + 0 + 4 + 16) + 2(9 + 1 + 1 + 9) \right),$$

$$S = \int_{-5}^{5} x^2 dx = 83.3333.$$

Main steps for solving the Simpson's 1/3 rule in MATLAB: For Simpson's rule two scripts are used. One for generating a general code and a second is used for any given function.

1) In the first script, the outputs and inputs are defined. S is the output which is the final value and the function 'func', initial value 'a', final value 'b' and the number of intervals 'n' are inputs.
2) In the formula, first and final functions are separated by using the values of a (initial) and b (final), i.e.,
   s = func(a),
   m = func(b).
3) The value of h is defined as
   h = (b − a)/n.
4) From the formula, even and odd parts will be eliminated separately. The first even part is calculated by starting the loop from 1:2:n − 1.
5) The domain x is written in the following form for the even interval, i.e.,
   x = a + h * i.
6) The values of x are picked by the function, i.e.,
   r = 4 * func(x).
7) To add these values, substitute t = 0 before the loop. And after writing the function t = t + r is implemented. This will add the entire even part.
8) Again the loop is implemented for the odd part from 2:2:n − 2.
9) The internal for odd function is
   x = a + h * i,
   r = 2 * func(x).
10) Before starting this loop, l = 0 is written (addition of all odd values).
11) After this function r, l = l + r is written for addition of odd terms.
12) At the end, the formula of the Simpson's 1/3 rule is written, i.e.,
    S = (h/3) * ((s + m) + t + l).
13) In the second script the function is defined. The function is saved by any name.
14) Now the inputs and outputs are implemented in the command window to get the desired result.

```
% Mfile 1
function S=Sympsons_One_third(func,a,b,n)
% S is the output
% func is the given function, a is the initial
% value, b is the final value and
% n is the number of intervals
s=func(a); % First value of function
m=func(b); % Final value of function
h=(b-a)/n; % Intervals
t=0; % For sequence to be added
for i=1:2:n-1
  x=a+h*i; % For even interval
  r=4*func(x); % Values of function for even
  t=t+r; % Addition of r
end
l=0; % For sequence to be added
for i=2:2:n-2
  x=a+h*i; % For odd interval
  r=2*func(x); % Values of function for odd
  l=l+r; % Addition of r
```

```
end
S= (h/3)*((s+m)+t+1); % Simpson 1/3 formula
```

```
% Mfile 2
function y=utv(x)
y=x^2;
end
```

```
>> % Command Window
>> S=Sympsons_One_third(@utv,-5,5,10)
S =
   83.3333
```

5.1.3 SIMPSON'S 3/8 RULE

In this method, the function in the integral $f(x)$ is replaced by a third order polynomial. For the determination of a third order polynomial on the curve, four points are needed (as shown in Figure 5.5). Consider the four points $(x_{i-1}, f(x_{i-1})), (x_i, f(x_i)), (x_{i+1}, f(x_{i+1}))$ and $(x_{i+2}, f(x_{i+2}))$ on the curve and assume that the polynomial passes through these points. The area under the curve is approximated by the area under the polynomial. Therefore, the polynomial is

$$P(x) = ax^3 + bx^2 + cx + d. \tag{5.28}$$

Similar to Simpson's 1/3 rule, the same procedure is applied to calculate the derivation of Simpson's 3/8 rule. The four points are substituted in Eq. (5.28) and we get four equations. These equations are solved, and the resulting area under the curve will be

$$S = \frac{3h}{8}\left(f(x_{i+2}) + 3f(x_i) + 3f(x_{i+1}) + f(x_{i-1})\right). \tag{5.29}$$

Eq. (5.29) is the integral for three segments, for the entire domain of equal width we have

$$S = \left(\frac{3h}{8}\right)\left(f(x_0) + 3\sum_{i=1,4,7...}^{n-2}\left(f(x_i) + f(x_{i+1})\right) + 2\sum_{i=3,6,9,...}^{n-3} f(x_i) + f(x_n)\right). \tag{5.30}$$

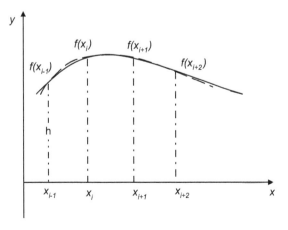

FIGURE 5.5 A graphical behavior of Simpson's 3/8 rule.

The procedure of calculating the truncation error of Simpson's 3/8 is similar to Simpson's 1/3. The truncation error per step is given by

$$T_r = -\frac{3}{80}h^5 f''''(\zeta_i), \text{ where } x_{i-1} < \zeta_i < x_{i+2}.$$ (5.31)

The total error is

$$TE = -\frac{1}{80}h^4(b-a)f'''_{avg}(\zeta_i).$$ (5.32)

Example 5.3: Evaluate $S = \int_{-5}^{5} x^2 dx$ by using Simpson's 3/8 rule.

Solution: Here $y(x) = x^2$. Assume h = 1. Hence:

x	−5	−4	−3	−2	−1	0	1	2	3	4	5
y	25	16	9	4	1	0	1	4	9	16	25

By formula

$$S = \int_{-5}^{5} x^2 dx = \frac{3}{8}\left((25+25)+3(16+9+1+0+4+9)+2(4+1+16)\right),$$

$$S = \int_{-5}^{5} x^2 dx = 78.3750.$$

Main steps for solving the Simpson's 3/8 rule in MATLAB: For Simpson's 3/8 rule two scripts are used. One for a general code and the second is used for any given function.

1) In the first script, the outputs and inputs are elaborated. S will be the output which is the final value, and the inputs are 'func' function, initial value 'a', final value 'b' and the number of intervals 'n'.
2) The first and final functions are separated by using the values of a (initial) and b (final).
   s = func(a),
   m = func(b).
3) Step size h is defined as:
   h = (b − a)/n.
4) In the second part of the formula, i.e., $(y_1 + y_2 + y_4 + ... + y_{n-1})$, the values which are multiples of three are missing. So, we must exclude them. The loop starts from 1:1:n − 1 with an increment 1 and a condition 'if' is implemented by restricting multiples of 3 as i~ = 3:3:n − 1.
5) The domain x is written in the following form:
   x = a + h * i.
6) The values of x are picked by the function
   r = func(x).
7) In order to add these values, t = 0 is implemented before the loop.
8) Again the loop is implemented for the multiples of 3 (by formula: $(y_3 + y_6 + ... + y_{n-2})$) from i = 3:3:n − 1.
9) For this the domain and the function are defined, i.e.,
   x = a + h * i,
   r = func(x).

10) The values of r are added by:
    t = t + r.
11) Before starting the loop for the third part of the formula I = 0 is written (addition).
12) A loop is implemented for a sequence which is a multiple of three and these values are substituted in the function r.
13) After this I = I + r is written for the addition of terms.
14) At the end, the formula of the Simpson's 3/8 rule is written, i.e.,
    S = (3 * h/8) * ((s + m) + 3 * t + 2 * I).
15) In the second script the function is defined.
16) Now the inputs and outputs are implemented in the command window to get the result.

```
% Mfile 1
function S=Sympsons_three_eight(func,a,b,n)
% S is the output
% func is the given function, a is the initial
% value, b is the final value and
% n is the number of intervals
s=func(a); % First value of the function
m=func(b); % Final value of the function
%(s+m) is the first part
h=(b-a)/n; % Intervals
t=0; % For a sequences to be added
for i=1:1:n-1
    if i~=3:3:n-1
    x=a+h*i; % Excluded values for multiples of 3
  r=func(x); % Values of the function for even
  t=t+r; % Addition of r (second part)
    end
end
l=0; % For a sequence to be added
for i=3:3:n-1
  x=a+h*i; % Multiples of 3
  r=func(x); % Values of function for multiples of 3
  l=l+r; % Addition of r
end
S=(3*h/8)*((s+m)+3*t+2*l); % Simpson 3/8 formula
```

```
% Mfile 2
function y=rst(x)
y=x^2;
end
```

```
>> % Command Window
>> S=Sympsons_three_eight(@rst,-5,5,10)
S =
   78.3750
```

5.2   RICHARDSON EXTRAPOLATION

Richardson extrapolation is used to improve the accuracy of the calculated results from a given numerical method provided that an estimated total error is given. First consider the case of trapezoidal rule. The total truncation error of the trapezoidal rule is $TE = T_e(h)^2$, where $T_e = -\dfrac{1}{12}(b-a)f''_{avg}$ (from Eq. 5.13). Assume that if T is the exact integral and T_1 and T_2 be the integral values calculated with different step sizes h_1 and h_2. From this we can write the total error as

$$T \approx T_1 + T_e\left(h_1\right)^2,\tag{5.33}$$

$$T \approx T_2 + T_e\left(h_2\right)^2.\tag{5.34}$$

From Eqs. (5.33) and (5.34) the value of T_e is eliminated as

$$T_e \approx \frac{T_2 - T_1}{\left(h_1\right)^2 - \left(h_2\right)^2}.\tag{5.35}$$

Substituting Eq. (5.35) into Eq. (5.34) yields

$$T \approx T_2 + \frac{T_2 - T_1}{\left(h_1 / h_2\right)^2 - 1}.\tag{5.36}$$

Eq. (5.36) gives an improved form of the integral as compared to the trapezoidal rule, the second term denotes the truncation error of integration having step size h_2. Assume that $h_1 / h_2 = 2$, then

$$T \approx \frac{4T_2 - T_1}{3}.\tag{5.37}$$

In a similar way, consider the Simpson's 1/3 rule. The truncation error for this rule is $TE = T_s\left(h\right)^4$ where $T_s = -\dfrac{1}{180}(b-a)f'''_{avg}(\zeta_i)$. By using the same procedure we have

$$T \approx T_1 + T_s\left(h_1\right)^4,\tag{5.38}$$

$$T \approx T_2 + T_s\left(h_2\right)^4.\tag{5.39}$$

Again, eliminating T_s and substituting the value in Eq. (5.39) gives

$$T \approx T_2 + \frac{T_2 - T_1}{\left(h_1 / h_2\right)^4 - 1}.\tag{5.40}$$

Assume $h_1/h_2 = 2$, then we have

$$T \approx \frac{16T_2 - T_1}{15}.\tag{5.41}$$

The obtained Eq. (5.41) is the improved form as compared to the Simpson's 1/3 rule.

5.2.1   ROMBERG INTEGRATION

As mentioned above, the Richardson extrapolation method is used to calculate the improved values of the integral. For better numerical results, this technique can be applied to calculate the higher order terms in the truncation error which is known as Romberg integration. If the higher order terms are added, then the total error (TE) in the trapezoidal rule is

$$TE = a_1\left(h\right)^2 + a_2\left(h\right)^4 + a_3\left(h\right)^6 \dots \tag{5.42}$$

Where a's denote constants. The first term on the right side was eliminated by the Richardson extrapolation method as calculated in the previous section. Assume that for i^{th} segment, the integral computed by the trapezoidal rule is denoted by $T_{i,1}$ and the integral calculated for the Richardson extrapolation is denoted by $T_{i,2}$, then from Eq. (5.37) we have

$$T_{i,2} \approx \frac{4T_{i,1} - T_{i-1,1}}{4^1 - 1}. \tag{5.43}$$

Again, apply the Richardson extrapolation for the second term and we will get $T_{i,3}$ which is

$$T_{i,3} \approx \frac{4^2 T_{i,2} - T_{i-1,2}}{4^2 - 1}. \tag{5.44}$$

The same procedure is repeated for the improved values of the integral. Generally, we can write

$$T_{i,j+1} \approx \frac{4^j T_{i,j} - T_{i-1,j}}{4^j - 1}, \tag{5.45}$$

or

$$T_{i+1,j+1} = T_{i+1,j} + \frac{\left(T_{i+1,j} - T_{i,j}\right)}{4^j - 1}. \tag{5.46}$$

In tabular form, the results are

TABLE 5.1
The General Form of Romberg Integration

$O(h^2)$	$O(h^4)$	$O(h^6)$	$O(h^{2n})$
$T_{1,1}$			
$T_{2,1}$	$T_{2,2}$		
$T_{3,1}$	$T_{3,2}$	$T_{3,3}$	
	$T_{4,2}$	$T_{4,3}$	. .
.	.	.	. .
.	.	.	. .
$T_{n,1}$	$T_{n,2}$	$T_{n,3}$	$T_{n,n}$

Example 5.4: Consider $T = \int_0^8 x^3 dx$, calculate the integral by using Romberg's method.

Solution: For $h = \frac{b-a}{2} = \frac{8-0}{2} = 4$ and $x = a + h$.

$$T_{1,1} = \int_0^8 x^3 dx = \frac{h}{2}\left[T_0 + 2T_1 + T_2\right] = \frac{4}{2}\left[0 + 2(64) + 512\right] = 1280.$$

For h = h/2 = 2,

$$T_{2,1} = \int_0^8 x^3 dx = \frac{h}{2}\left[T_0 + 2(T_1 + T_2 + T_3) + T_4\right] = \frac{2}{2}\left[0 + 2(8 + 64 + 216) + 512\right] = 1088.$$

Similarly, for h = h/2 = 1,

$$T_{3,1} = \int_0^8 x^3 dx = \frac{h}{2}\left[T_0 + 2\left(T_1 + T_2 + T_3 \ldots T_7\right) + T_8\right] = 1040.$$

For h = h/2 = 0.5,

$$T_{4,1} = \int_0^8 x^3 dx = \frac{h}{2}\left[T_0 + 2\left(T_1 + T_2 + T_3 \ldots T_{15}\right) + T_{16}\right] = 1028.$$

The $O(h^4)$ approximation yields

$$T_{2,2} = T_{2,1} + \frac{\left(T_{2,1} - T_{1,1}\right)}{3} = 1024,$$

$$T_{3,2} = T_{3,1} + \frac{\left(T_{3,1} - T_{2,1}\right)}{3} = 1024,$$

$$T_{4,2} = T_{4,1} + \frac{\left(T_{4,1} - T_{3,1}\right)}{3} = 1024.$$

The $O(h^6)$ approximation gives

$$T_{3,3} = T_{3,2} + \frac{\left(T_{3,2} - T_{2,2}\right)}{3} = 1024,$$

$$T_{4,3} = T_{4,2} + \frac{\left(T_{4,2} - T_{3,2}\right)}{3} = 1024.$$

The $O(h^8)$ approximation gives

$$T_{4,4} = T_{4,3} + \frac{\left(T_{4,3} - T_{3,3}\right)}{3} = 1024.$$

In tabular form, the results are:

$O(h^2)$	$O(h^4)$	$O(h^6)$	$O(h^8)$
1280			
1088	1024		
1040	1024	1024	
1028	1024	1024	1024

Main steps for solving by using Romberg's method in MATLAB: In this method, the required table T is output. The inputs are the function 'func' which is calculated in the second script, starting point 'a', ending point 'b' and the number of rows and columns which are denoted by n.

1) h is defined next as h = (b − a)/2. Where 'a' and 'b' are given.
2) The results are in matrix form, for this a n × n matrix is defined in the form of 'zeros', i.e.,
   T = zeros(n).
3) Before applying the trapezoidal rule, x is defined with increment h as
   x = x + h.
4) Next apply the trapezoidal rule, i.e.,
   T(1,1) = (h/2) * (func(a) + 2 * func(x) + func(b)).
Note that T(1,1) denotes the first entry of the matrix.
5) For T(2,1), we reduce h by h/2.

6) For adding all the terms appearing in the middle function, s = 0 is written before the loop.

7) For reduced h the loop starts from 1 to 3, x is defined as
   x = a + h * (i).

8) The value of x is used in the function func(x) along with the addition of s, i.e.,
   s = s + func(x).
   The s will add all values.

9) T(2,1) is defined as:
   T(2,1) = (h/2) * (func(a) + 2 * s + func(b)).
   This will give us the second entry below the first one.

10) The same procedure is repeated again by reducing h and the loop starts from 1 to 7 we will get T(3,1).
    T(3,1) = (h/2) * (func(a) + 2 * t + func(b)).

11) The same process is used for T(4,1) but the loop starts from 1 to 15.

12) We will get a single column in the form, i.e.,

$$
\begin{array}{c}
\hline
T(1,1) \\
T(2,1) \\
T(3,1) \\
T(4,1) \\
\hline
\end{array}
$$

13) By using these values we will calculate T(2,2), T(3,2), T(4,2), T(3,3), T(4,3) and T(4,4).

14) For this the loop is implemented from j = 1:n – 1and i = j:n – 1 and the formula will be

$$T(i + 1, j + 1) = T(i + 1, j) + (T(i + 1, j) - T(i, j))/(4 \wedge j - 1).$$

This will give us a complete table.

15) The second script is used to write the function.

16) The command window is used to run both files.

```
% Mfile 1
function T=Rombergs(func,a,b,n)
% Initial and final values are a and b
% n is the n*n matrix
h=(b-a)/2;
T=zeros(n); % T is n*n matrix
x=a+h; % Increment in x
T(1,1)=(h/2)*(func(a)+2*func(x)+func(b));
% T(1,1)is the first entry of the trapezoidal rule
h=h/2; % Reducing h
s=0; % Adding sequence
for i=1:3
 x=a+h*(i);
 s=s+func(x); % Middle terms of the trapezoidal rule
end
T(2,1)=(h/2)*(func(a)+2*s+func(b));
h=h/2; % Reducing h
t=0; % Adding sequence
for i=1:7
 x=a+h*(i);
 t=t+func(x); % Middle terms of the trapezoidal rule
end
T(3,1)=(h/2)*(func(a)+2*t+func(b));
```

```
h=h/2; % Reducing h
l=0;
for i=1:15
  x=a+h*(i);
  l=l+func(x); % Middle terms of the trapezoidal rule
end
T(4,1)=(h/2)*(func(a)+2*l+func(b));
% The remaining entries are calculated by using these
%values
for j=1:n-1
  i=j:n-1
  T(i+1,j+1)=T(i+1,j)+(T(i+1,j)-T(i,j))/(4^j-1);
end
```

```
% Mfile 2
function y=rst(x)
y=x^3;
end
```

```
>> % Command window
>> Rombergs(@rst,0,8,4)
ans =
      1280         0         0         0
      1088      1024         0         0
      1040      1024      1024         0
      1028      1024      1024      1024
```

5.3   ADAPTIVE QUADRATURE

In this chapter, the methods we have studied so far are used to estimate the integral of a given function for equally spaced points. There are several functions which show abrupt change in some regions (as illustrated in Figure 5.6). A more refinement in the step size is needed in those regions. Adaptive quadrature methods have grip to tackle in such regions and automatically adjust the step size so that small steps are considered in such regions and large steps are considered where functions change slowly. In this technique the composite Simpson's 1/3 rule is applied at two levels of

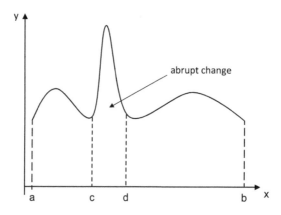

FIGURE 5.6   A graphical behavior of abrupt change in the function.

improvement and the difference between these two levels is used to evaluate the truncation error. If the truncation error is small, then there is no need to improve the step size but if the error is large then the step size is modified, and the procedure is repeated until the error falls. Theoretically, consider the interval [a,b] with a width $h_1 = b - a$. By using the Simpson's 1/3 rule

$$S(h_1) = \frac{h_1}{6}\big(f(a) + 4f(c) + f(b)\big), \tag{5.47}$$

where $c = (a+b)/2$.

By using the Richardson extrapolation method, a more improved result can be calculated by halfling the step size. By using the Simpson's 1/3 rule with four subintervals:

$$S(h_2) = \frac{h_2}{6}\big(f(a) + 4f(d) + f(c)\big) + \frac{h_2}{6}\big(f(c) + 4f(e) + f(b)\big), \tag{5.48}$$

where $h_2 = h_1/2$, $d = (a+c)/2$ and $e = (c+b)/2$.

The error is

$$E \cong S(h_2) - S(h_1). \tag{5.49}$$

Using the approach of the Richardson extrapolation method, we will get a more modified result, i.e.,

$$E(h_2) = \frac{1}{15}\big(S(h_2) - S(h_1)\big). \tag{5.50}$$

If the error is not less than given tolerance, then the interval is further subdivided into equal subintervals and the same procedure is repeated until we get the condition satisfied.

Example 5.5: Consider $S = \int_{-5}^{5} x^2 dx$, calculate the integral by using the adaptive quadrature with tolerance 0.001.

Solution: First use the Simpson's rule with two subintervals to obtain an approximate result of this integral:

$$S_1 = \frac{10}{6}\big(25 + 4(0) + 25\big) = 83.3333.$$

Now, divide the interval $[-5,5]$ into two subintervals of equal width, $[-5,0]$ and $[0,5]$, and apply the Simpson's rule to get the second approximation S_2:

$$S_2 = \frac{5}{6}\big(25 + 4(6.2500) + 0\big) + \frac{5}{6}\big(0 + 4(6.2500) + 25\big) = 83.3333.$$

Calculating the error yields

$$E(h_2) = \frac{1}{15}\big(83.3333 - 83.3333\big) = 0 < 0.001.$$

Main steps for solving the adaptive quadrature rule in MATLAB: In this method, S is the output. The inputs are the function 'func' which is calculated in the second script and the starting and end points 'a' and 'b'.

1) The Simpson's rule for two subintervals is calculated. The initial and final points are given, the middle point is obtained by taking the average of initial and final points, i.e., c = (a + b)/2.
2) The value of h is calculated as
   h = b − a.
3) The formula for Simpson's rule is implemented, i.e.,
   s1 = (h1/6) * (func(a) + 4 * func(c) + func(b)).
4) Next, we will half the step size (h/2) and apply the Simpson's formula for four points. But before writing the formula in MATLAB the two unknown values are defined:
   d = (a + c)/2,
   e = (c + b)/2.
5) The formula for four points is written:
   s2 = h2/6 * (func(a) + 4 * func(d) + func(c)) + h2/6 * (func(c) + 4 * func(e) + func(b)).
6) The error is calculated by
   E = (1/15) * (s2 − s1).
7) The second script is used to write the given function and the command window is used to run both files.

```
% Mfile 1
function E=adaptive(func,a,b)
% S is the output
% func is the given function, a and b are the given domain
c=(a+b)/2;
h1=(b-a);
s1=(h1/6)*(func(a)+4*func(c)+func(b));
h2=h1/2;
d=(a+c)/2;
e=(c+b)/2;
s2=h2/6*(func(a)+4*func(d)+func(c))+h2/6*(func(c)+4*func(e)+func
(b));
E=(1/15)*(s2-s1);
end
```

```
% Mfile 2
function y=rsm(x)
y=x^2;
end
```

```
>> % Command Window
>> adaptive(@rsm,-5,5)
ans =
    0
```

5.4 GAUSSIAN QUADRATURE

To calculate the value of $\int_a^b f(x)\,dx$, various numerical methods are used to approximate $f(x)$ by using a polynomial, along with fixed and equally spaced points. But if the points are not fixed, we have the option to pick any point in such a way that the error can be minimized. For example, consider the case of trapezoidal rule (as shown in Figure 5.7a) in which the fixed point on the curve is a and b. But if we have the option to choose any two points on the curve then we can choose in such a

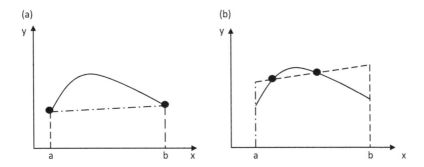

FIGURE 5.7   (a) Graphical behavior of trapezoidal rule and (b) Graphical behavior of Gauss quadrature.

way that the resulting area of the trapezoid gives a better estimate (as illustrated in Figure 5.7b). The Gaussian quadrature relies on this approach, it approximates the integral by taking the weighted sum of $f(x)$ at several points which are not fixed in $[a,b]$.

Thus, the Gauss quadrature depends on the variety of interpolating functions which give maximum accuracy. Hence the two points of the integral $Q = \int_{-1}^{1} f(\varsigma) d\varsigma$ is approximated by

$$Q = q_1 f(\varsigma_1) + q_2 f(\varsigma_2). \tag{5.51}$$

Where $\varsigma_1, \varsigma_2, q_1$ and q_2 are unknowns. The limits a and b are converted into the limits 1 and -1 by using the transformation:

$$x = a + \frac{(b-a)(\varsigma+1)}{2}. \tag{5.52}$$

To calculate the four unknowns in Eq. (5.51) we adapt the method of undetermined coefficients. Since there are four unknowns, we assume that the method is correct for four functions 1, x, x^2 and x^3. Substitution gives

$$q_1 + q_2 = \int_{-1}^{1} 1 d\varsigma = 2,$$

$$q_1 \varsigma_1 + q_2 \varsigma_2 = \int_{-1}^{1} \varsigma d\varsigma = 0,$$

$$q_1 \varsigma_1^2 + q_2 \varsigma_2^2 = \int_{-1}^{1} \varsigma^2 d\varsigma = \frac{2}{3},$$

$$q_1 \varsigma_1^3 + q_2 \varsigma_2^3 = \int_{-1}^{1} \varsigma^3 d\varsigma = 0. \tag{5.53}$$

Solving gives

$$\varsigma_1 = -\frac{1}{\sqrt{3}} = -0.57735, \varsigma_2 = \frac{1}{\sqrt{3}} = 0.57735, q_1 = 1, q_2 = 1.$$

Hence Eq. (5.51) becomes:

$$Q = f\left(-\frac{1}{\sqrt{3}}\right) + f\left(\frac{1}{\sqrt{3}}\right). \tag{5.54}$$

TABLE 5.2

Points and Weights used in the Gaussian Quadrature

n	Weights q_i	Points ς_i
2	$q_1 = 1$	$\varsigma_1 = -0.57735$
	$q_2 = 1$	$\varsigma_2 = 0.57735$
3	$q_1 = 0.55555$	$\varsigma_1 = -0.77459$
	$q_2 = 0.88888$	$\varsigma_2 = 0$
	$q_3 = 0.55555$	$\varsigma_3 = 0.77459$
4	$q_1 = 0.34785$	$\varsigma_1 = -0.86113$
	$q_2 = 0.65214$	$\varsigma_2 = -0.33998$
	$q_3 = 0.65214$	$\varsigma_3 = 0.339981$
	$q_4 = 0.34785$	$\varsigma_4 = 0.861136$
5	$q_1 = 0.23693$	$\varsigma_1 = -0.90618$
	$q_2 = 0.47863$	$\varsigma_2 = -0.53847$
	$q_3 = 0.56889$	$\varsigma_3 = 0$
	$q_4 = 0.47863$	$\varsigma_4 = 0.53847$
	$q_5 = 0.23693$	$\varsigma_5 = 0.90618$

The Eq. (5.54) is known as the Gauss Legendre formula for two points (n = 2). The results are calculated for two function evaluations, the number of functions can be increased for more accurate results (as shown in Table 5.2). The general form of this method can be written as:

$$Q = q_1 f(\varsigma_1) + q_2 f(\varsigma_2) + \ldots q_n f(\varsigma_n). \tag{5.55}$$

The error for n-point formula is [4–80]

$$E \approx \frac{2^{2n+1}(n!)^4}{(2n+1)(2n!)^3} f^{2n}(\varsigma), \text{where} -1 < \varsigma < 1. \tag{5.56}$$

Example 5.6: Use the two-point Gauss Legendre formula to evaluate the following function:

$$f(x) = x^3 e^x,$$

between the limits x = 0 to 3.
Solution: Firstly, changing the limits from –1 to 1 and substituting a = 0 and b = 3 into Eq. (5.52)

$$x = a + \frac{(b-a)(\varsigma+1)}{2} = 1.5 + 1.5\varsigma \text{ and } dx = 1.5d\varsigma.$$

Substituting these values in the original equation

$$Q = \int_0^3 (x^3 e^x) dx = 1.5 \int_{-1}^1 \left[(1.5 + 1.5\varsigma)^3 e^{1.5+1.5\varsigma} \right] d\varsigma,$$

$$Q = 1.5 \left[\left(\left(1.5+1.5(-0.57735)\right)^3 e^{1.5+1.5(-0.57735)} \right) + \left(\left(1.5+1.5(0.57735)\right)^3 e^{1.5+1.5(0.57735)} \right) \right]$$

$$= 212.4108.$$

Main steps for solving the Gaussian quadrature rule in MATLAB: In this method, G is the output. The inputs are the function 'func' which is calculated in the second script, the starting point 'a', end point 'b' and the number of points 'n'.

1) The values of the weights q_i's and points ς_i's are defined by using the 'if' command.
2) Now for addition and multiplication of a sequence we will define s = 0 and xt = 1 before implementing the loop.
3) The loop is started, define $x = a + \dfrac{(b-a)(\varsigma+1)}{2}$ and this value is used in the function which can be written as:
   g = func(d).
4) The formula $Q = q_1 f(\varsigma_1) + q_2 f(\varsigma_2) + \ldots q_n f(\varsigma_n)$ can be written as:
   s = s + q(i) * g.
Before starting the loop, we define s = 0, this procedure is used for the addition of the sequence.
5) At the end the value of s is used in the formula
   G = (b − a) * s/2.
6) The second script is used to write the given function and the command window is used to run both files.

```
% Mfile 1
function G=Gausss_quadrature(func,a,b,n)
if n==2
  x(1)=-0.57735; x(2)=0.57735;
  q(1)=1; q(2)=1;
end
if n==3
  x(1)=-0.77459; x(2)=0; x(3)=0.77459;
  q(1)=0.55555; q(2)=0.88888;q(3)=0.55555;
end
if n==4
  x(1)=-0.86113; x(2)=-0.33998; x(3)=0.339981;
  x(4)=0.861136;
  q(1)=0.34785; q(2)=0.65214;q(3)=0.65214;
  q(4)=0.34785;
end
if n==5
  x(1)=-0.90618; x(2)=-0.53847; x(3)=0;
  x(4)=0.53847;x(5)=0.90618;
  q(1)=0.23693; q(2)=0.47863;q(3)=0.56889;
  q(4)=0.47863;q(5)=0.23693;
end
s=0;
xt=1;
for i=1:n
  d=((b-a)*x(i)+a+b)/2;
g=func(d);
xt=xt*q(i);
s=s+q(i)*g;
```

```
end
G=(b-a)*s/2;
```

```
% Mfile 2
function y=adf(x)
y=x^3*exp(x);
end
```

```
    >> % Command Window
>> Gausss_quadrature(@adf,0,3,2)
ans =
  212.4108
```

PROBLEM SET 5

1. Using the Mfile, evaluate the following integral:

$$\int_0^2 \frac{dx}{1+x+x^2}.$$

By using (a) single trapezoid, (b) composite trapezoidal rule with $n = 1$ and 2, (c) single Simpson's 1/3 rule, (d) composite Simpson's 1/3 rule with n = 2, (e) Simpson's 3/8 rule.

2. Develop a code for the following:

$$\int_0^\pi e^{\sin(x)} dx.$$

By using (a) single trapezoid, (b) composite trapezoidal rule with $n = \pi$, (c) single Simpson's 1/3 rule, (d) composite Simpson's 1/3 rule with $n = \pi$, (e) Simpson's 3/8 rule.

3. Calculate the temperature at a certain depth of earth over a period. Suppose that the earth is at zero temperature and flat. At depth d the temperature $T(d,t)$ is taken to be positive at time t, which is:

$$T(d,t) = \frac{d}{2b} \int_0^t \frac{e^{-\frac{d^2}{4b(t-\sigma)}}}{(t-\sigma)\sqrt{4\pi b(t-\sigma)}} T_q(\sigma) d\sigma.$$

Temperature at the surface of the earth is denoted by $T_q(\sigma)$ and b is the thermal diffusivity. Assume t in hours then

$$T_q(\sigma) = 20\sin\left(\frac{2\pi t}{8766}\right) + 15.$$

Assume $b = 0.01 m^2/hr$. Use Mfile to calculate $T(d,t)$ for the following values: $t = 100, 200, 300$ and $d = 2\,m$.

4. Assume

$$\int_0^2 \frac{xdx}{1+x^2}.$$

Use (a) composite trapezoidal rule, (b) composite Simpson's 1/3 rule and (c) Simpson's 3/8 rule to calculate the integral by using Mfile.

5. Determine the velocity v of a particle moving in a straight line that covers a distance x in time t. The values are:

x	0	10	20	30	40
V	40	55	60	49	36

Calculate time taken to cover the distance of 40 units.

6. Using Mfile, apply (a) composite trapezoidal rule, (b) composite Simpson's 1/3 rule and (c) Simpson's 3/8 rule to the data in the table:

x	0	1	1.1	1.2	1.3	1.4	1.5	1.6	1.7	1.8	1.9	2
y	0	0.36	0.28	0.24	0.19	0.15	0.9	0.01	0.04	0.03	0.02	0.01

7. Using Mfile, apply (a) composite trapezoidal rule, (b) composite Simpson's 1/3 rule and (c) Simpson's 3/8 rule to the following integral: $\int_0^1 \dfrac{dx}{2 + cos(5\pi x)}$.

8. Use Mfile and calculate the integral by using Romberg's method:

$$\int_0^3 sin\left(x^2 cos\left(e^{-x}\right)\right)e^{-x}dx.$$

9. Generate a code of the following function by using Romberg's method:

$$\int_0^\pi cos\left(3sin(2x) + 3cos(3x) + 2cos(2x) + 3sin(x) + cos(x)\right)dx.$$

10. Use Romberg's method in MATLAB to calculate the following integral:

$$\int_{-1}^3 \left(2x^5 - 4x^3 - x + 2\right)dx.$$

11. Using MATLAB, evaluate the integral:

$$\int_{1.1}^{2.1} \left(2x^2 + \frac{3}{x}\right)^2 dx.$$

By using (a) adaptive quadrature and (b) Gauss quadrature methods.

12. Use MATLAB to evaluate the integral:

$$\int_1^9 \left(-0.04x^4 + 0.872x^3 - 4.123x^2 + 5.21x + 3\right)dx.$$

By using (a) adaptive quadrature and (b) Gauss quadrature methods.

13. Figure 5.8 illustrates a uniform intensity p carried by a circle of radius b. The vertical displacement of the surface is shown by point q, we have

$$f(r) = f_0 \int_0^{\pi/2} \frac{cos^2(\theta)d\theta}{\sqrt{\left(\frac{r}{b}\right)^2 - sin^2(\theta)}},$$

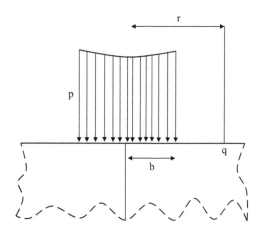

FIGURE 5.8 Uniform intensity p carried by a circle.

where f_0 denotes the displacement at $r = b$. Use (a) adaptive quadrature and (b) Gauss quadrature codes to calculate the integral at $r = 2b$.

6 Solution of Initial Value Problems (IVPs)

Differential equations are widely used for mathematical modeling in engineering and science. Many physical phenomena are analyzed by differential equations, as the rate of change of a physical quantity can be expressed in the form of a differential equation. An equation which involves at least one derivative of the unknown function is called a differential equation. In this chapter, our focus will be on the equation of the form

$$\frac{dy}{dx} = f(x, y), \tag{6.1}$$

with initial condition

$$y(x_0) = y_0. \tag{6.2}$$

Eq. (6.1) along with condition (6.2) is called the initial value problem. If a thin pin is dropped from some point y_0 (as shown in Figure 6.1), this phenomenon is analyzed by Newton's second law, i.e., $F = ma$. As the acceleration a is the rate of the change of the velocity with respect to time. Moreover, another force in the form of air resistance is also included, i.e., $F = kv$ (for small velocity). Where k depends on the shape and size of the pin as well as the viscosity and density of the air. Hence the differential equation along with the initial condition becomes

$$m\frac{dv}{dt} = -mg - kv, \tag{6.3}$$

$$y(t_0) = y_0. \tag{6.4}$$

In most of the problems, the analytical solution is difficult to calculate, and in such circumstances numerical approximations are mostly preferred. There are two types of methods that will be studied in this chapter, namely, single step and multistep methods.

FIGURE 6.1 A graphical behavior of a thin pin dropped from some point.

DOI: 10.1201/9781003385288-6

6.1 SINGLE STEP METHODS

In single step methods, the slope estimated at the previous station y_i located at x_i is used to extrapolate the solution at the station y_{i+1} located at x_{i+1}. Figure 6.2 shows the graphical illustration of single step methods, where $\varnothing$ is the slope that estimates the extrapolation from the previous station y_i to the current station y_{i+1}, i.e.,

$$y_{i+1} = y_i + h\varnothing, \quad i = 0,1,2,\ldots n-1. \tag{6.5}$$

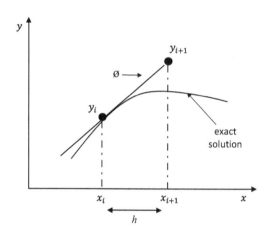

FIGURE 6.2 A graphical illustration of a single step method.

6.1.1 EULER'S METHOD

The first method, known as Euler's method, helps to calculate the approximate solution of the initial value problem. This method has limited use due to its large error which is stored as the procedure proceeds.

Consider $\dfrac{dy}{dx} = f(x,y)$ with initial condition $y(x_0) = y_0$. Let [a,b] be the interval from which we want to calculate the solution of the problem. We subdivide the interval [a,b] into equal subintervals, i.e., $x_i = x_0 + ih$, $i = 0,1,2,\ldots$.

The expansion of $y(x_1)$ in the Taylor's series about point $x = x_0$ is

$$y(x_1) = y(x_0) + y'(x_0)(x_1 - x_0) + y''(x_0)\frac{(x_1 - x_0)^2}{2}\ldots \tag{6.6}$$

Substitute $y'(x_0) = f(x_0, y(x_0))$ and $h = x_1 - x_0$ in Eq. (6.6) and neglecting the second order terms and above power terms yields

$$y_1 = y_0 + hf(x_0, y_0). \tag{6.7}$$

The above expression is required form of Euler's formula.

The general form is

$$y_{i+1} = y_i + hf(x_i, y_i), \text{ where } x_{i+1} = x_i + h, \text{ for } i = 0,1,2,\ldots n-1. \tag{6.8}$$

As can be seen in Eq. (6.8) and Eq. (6.5) the only difference is that the slope $\varnothing$ is replaced with $f(x_i, y_i)$, which is the slope $y'(x_i)$. The graphical interpretation of Euler's method is the same as mentioned in Figure 6.2.

6.1.1.1 Error Analysis

Two types of errors are involved in calculating the solution of ordinary differential equations: roundoff and truncation. Roundoff occurs due to the reduced values of the significant digits. While the truncation error consists of two parts: Local truncation error and global truncation error. The numerical procedure performed in a single step from x_i to x_{i+1} involves a local truncation error. The addition of the truncated errors (each step) is called the global truncation error. In order to calculate the global error for Euler's method assume that the global error after n steps be

$$G_n = y(x_n) - y_n, \text{where } n = 0,1,2\ldots. \tag{6.9}$$

Where $y(x_n)$ denotes the true solution and y_n is the approximate solution at $x = x_n$. Eq. (6.6) can be written as

$$y(x_{n+1}) = y(x_n) + f(x_n, y(x_n))h + y'' \ (\zeta_n)\frac{h^2}{2}, \quad x_n < \zeta_n < x_{n+1}. \tag{6.10}$$

Subtract Eq. (6.10) and Eq. (6.8) and we get

$$y(x_{n+1}) - y_{n+1} = y(x_n) - y_n + h(f(x_n, y(x_n)) - f(x_n, y_n)) + y''(\zeta_n)\frac{h^2}{2}, \tag{6.11}$$

using Eq. (6.9) in Eq. (6.11)

$$G_{n+1} = G_n + h(f(x_n, y(x_n)) - f(x_n, y_n)) + y''(\zeta_n)\frac{h^2}{2},$$

$$= G_n + h(\frac{f(x_n, y(x_n)) - f(x_n, y_n)}{(y(x_n) - y_n)})(y(x_n) - y_n) + y''(\zeta_n)\frac{h^2}{2},$$

$$= G_n + h(\frac{f(x_n, y(x_n)) - f(x_n, y_n)}{(y(x_n) - y_n)})G_n + y''(\zeta_n)\frac{h^2}{2}. \tag{6.12}$$

Assume that the function is continuous, using the mean value theorem

$$G_{n+1} = G_n + hf_y(x_n, \rho_n)G_n + y''(\zeta_n)\frac{h^2}{2}, \quad y_n < \rho_n < y(x_n). \tag{6.13}$$

Assume that $|f_y(x_n, \rho_n)| \leq N$ and $|y''(\zeta_n)| \leq F$, where N and F are positive constants, so Eq. (6.13) becomes

$$|G_{n+1}| < (1 + hN)|G_n| + F\frac{h^2}{2}. \tag{6.14}$$

Now we will show by induction that

$$|G_n| \leq d((1 + hN)^n - 1), \quad n = 0,1,2\ldots \tag{6.15}$$

Where $d = \dfrac{hF}{2N}$.

Since $y(x_0) = y_0$, then Eq. (6.15) is true for $n = 0$. Suppose that Eq. (6.15) is true for an integer n, then we will show that it is true for n + 1. Hence

$$|G_{n+1}| \le d\left((1+hN)^{n+1} - 1\right). \tag{6.16}$$

Substituting Eq. (6.15) in Eq. (6.14)

$$|G_{n+1}| \le (1+hN)d\left((1+hN)^n - 1\right) + F\frac{h^2}{2},$$

$$= d\left((1+hN)^{n+1} - (1+hN)\right) + F\frac{h^2}{2}.$$

As $hNd = F\dfrac{h^2}{2}$, hence it will satisfy Eq. (6.15).

We know that $1 + x \le e^x$ for $x \ge 0$. From this relation we can write

$$(1+hN) \le e^{hN},$$

$$(1+hN)^n \le e^{nhN}. \tag{6.17}$$

Using Eq. (6.17) in Eq. (6.15)

$$|G_n| \le F\frac{h}{2N}\left(e^{nhN} - 1\right). \tag{6.18}$$

Eq. (6.18) shows that the Euler method has a global error of order h.

Example 6.1: Use Euler's method to find the numerical solution of the problem $\dfrac{dy}{dx} = xy$, $y(0) = 1$ for $x = 0, 0.2, 0.4, 0.6, 0.8$ and 1.

Solution: Here $h = 0.2, f(x,y) = xy, x_0 = 0, y_0 = 1, x_1 = 0.2, x_2 = 0.4, x_3 = 0.6, x_4 = 0.8$ and $x_5 = 1$. By Euler's formula, for i = 0

$$y_1 = y_0 + hf(x_0, y_0) = y_0 + hx_0y_0 = 1 + (0.2)(0)(1) = 1,$$

for i = 1

$$y_2 = y_1 + hf(x_1, y_1) = y_1 + hx_1y_1 = 1 + (0.2)(0.2)(1) = 1.0400,$$

for i = 2, 3 and 4 the results will be

$$y_3 = 1.1232, y_4 = 1.2580 \text{ and } y_5 = 1.4593.$$

Main steps for solving the Euler's method in MATLAB: The required values of y will be output (denoted by z). The inputs will be function 'func' (which will be implemented in the second script), y0, x0 (initial values given), increment h and the final value xf.

1. Now x is defined by the colon operator
   x = x0:h:xf.
2. Before starting the loop, the value y0 is represented as
   y(1) = y0.
   Note: By using this, the value will be used in a loop. Simply using y0 can't be used for calling the value y0 in a loop.
3. The loop starts from
   i = 1:length(x) – 1.
   Note: For making equal dimensions –1 is subtracted. As y0 is already defined before the loop, so 1 must be subtracted in order to make x and y having equal dimensions.
4. The Euler's formula is implemented
   y(i + 1) = y(i) + h * func(x(i),y(i)).
5. And the loop is ended.
6. At the end replace z = y as the output will be in the form z.
7. Now the second script is used to write the given function.
8. The two scripts are used to implement the results on the command window.

```
% Mfile 1
function [z]=euler_Method(func,y0,x0,h,xf)
% z is the output
% func, y0, x0,h and xf are the inputs
x=x0:h:xf; % Domain
y(1)=y0; % The initial value is used in the loop
for i=1:length(x)-1 % In order to make equal dimensions
    y(i+1)=y(i)+h*func(x(i),y(i)); % Euler's formula
end
z=y;
```

```
% Mfile 2
function [z]=tys(x,y)
z=x*y;
end
```

```
>> % Command Window
>> euler_Method(@tys,1,0,0.2,1)
    ans =
         1.0000  1.0000  1.0400  1.1232  1.2580  1.4593
```

6.1.1.2   Euler's Method for the Systems of Ordinary Differential Equations

To solve

$$y'' = f(x, y, y') \text{ with } y(x_0) = y_0, y'(x_0) = y_0',$$ (6.19)

set

$$y' = z, y'' = z'.$$ (6.20)

By substituting Eq. (6.20) into Eq. (6.19), the differential equation reduces to the first order system

$$y' = z,$$

$$z' = f(x, y, z). \tag{6.21}$$

The conditions become

$$y(x_0) = y_0, z(x_0) = z_0. \tag{6.22}$$

Note: In this chapter, this procedure will be used for all the systems of equations.

Example 6.2: Solve $y'' + 3y' - 2y = 5$ with $y(0) = 0$, $y'(0) = 1$ and h = 0.1, and calculate $y(1)$.

Solution: Let

$$z = y',$$

So

$$z' = 5 - 3z + 2y,$$

with initial conditions reduced to

$$y(0) = 0, z(0) = 1.$$

We have $h = 0.1, x_0 = 0, y_0 = 0, z_0 = 1$.

Since $x_0 = 0, x_1 = x_0 + h = 0.1,$

$x_2 = x_0 + 2h = 0.2, x_3 = x_0 + 3h = 0.3,$

$x_4 = x_0 + 4h = 0.4, x_5 = x_0 + 5h = 0.5,$

$x_6 = x_0 + 6h = 0.6, x_7 = x_0 + 7h = 0.7,$

$x_8 = x_0 + 8h = 0.8, x_9 = x_0 + 9h = 0.9,$

$x_{10} = x_0 + 10h = 1.$

Using Euler's formula for i = 0

$$y_1 = y_0 + hf(x_0, y_0, z_0) = y_0 + h(z_0) = 0 + (0.1)(1) = 0.1000,$$

$$z_1 = z_0 + hg(x_0, y_0, z_0) = z_0 + h(5 - 3z_0 + 2y_0) = 1 + 0.1(5 - 3(1) + 2(0)) = 1.2000,$$

for i = 1

$$y_2 = y_1 + hf(x_1, y_1, z_1) = y_1 + h(z_1) = 0.1 + 0.1(1.2) = 0.2200,$$

$$z_2 = z_1 + hg(x_1, y_1, z_1) = z_1 + h(5 - 3z_1 + 2y_1) = 1.2 + 0.1(5 - 3(1.2) + 2(0.1)) = 1.3600,$$

for i = 2, 3, 4, 5, 6, 7, 8 and 9, the results will be

$$\begin{pmatrix} y_3 \\ z_3 \end{pmatrix} = \begin{pmatrix} 0.3560 \\ 1.4960 \end{pmatrix}, \begin{pmatrix} y_4 \\ z_4 \end{pmatrix} = \begin{pmatrix} 0.5056 \\ 1.6184 \end{pmatrix}, \begin{pmatrix} y_5 \\ z_5 \end{pmatrix} = \begin{pmatrix} 0.6674 \\ 1.7340 \end{pmatrix}, \begin{pmatrix} y_6 \\ z_6 \end{pmatrix} = \begin{pmatrix} 0.8408 \\ 1.8473 \end{pmatrix}, \begin{pmatrix} y_7 \\ z_7 \end{pmatrix} = \begin{pmatrix} 1.0256 \\ 1.9613 \end{pmatrix},$$

$$\begin{pmatrix} y_8 \\ z_8 \end{pmatrix} = \begin{pmatrix} 1.2217 \\ 2.0780 \end{pmatrix}, \begin{pmatrix} y_9 \\ z_9 \end{pmatrix} = \begin{pmatrix} 1.4295 \\ 2.1989 \end{pmatrix}, \begin{pmatrix} y_{10} \\ z_{10} \end{pmatrix} = \begin{pmatrix} 1.6494 \\ 2.3252 \end{pmatrix}.$$

Main steps for solving the systems of equations by Euler's method in MATLAB: The values of y and z will be output (denoted by z in MATLAB). The inputs will be function 'func' (which will be implemented in the second script), y1, x0 (initial values given), increment h and the final value xf.

1. The domain is defined by the colon operator, i.e.,
   x = x0:h:xf.
2. As we have two values, $y_0 = 0$ and $z_0 = 1$. this phenomenon can be written as
   y(:,1) = y1.
   Note that this command is used for one column and rows will depend on the order of the equation. If we have a third order equation, then three rows will appear and so on. In this case both values 0 and 1 are in the form:
   0
   1
3. Now the loop is started from:
   i = 1:length(x) − 1.
4. The main step here is how to write Euler's formula for higher order equations. As there are two equations we will deal with two rows, so we can write:
   y(:,i + 1) = y(:,i) + h * func(x(i),y(:,i)).
   This will represent two rows and one column.
5. The loop is ended in the next step.
6. The second script is used to write the given second order equation in the form:
   z = [y(2); 5 − 3 * y(2) + 2 * y(1)].
7. Now the command window is used to get the desired result.

```
% Mfile 1
function [z]=euler_Method_System(func,y1,x0,h,xf)
% z is the output
% func, y1, x0,h and xf are the inputs
x=x0:h:xf; % Domain
y(:,1)=y1; % One column
for i=1:length(x)-1
  y(:,i+1)=y(:,i)+h*func(x(i),y(:,i)); % Euler's formula
end
z=y'; % Column vectors


% Mfile 2
function [z]=tys1(x,y)
z=[y(2); 5-3*y(2)+2*y(1)];
end


>> % Command Window
>> [z]=euler_Method_System(@tys1,[0 1],0,0.1,1)
```

6.1.2 HEUN'S METHOD

The difference between Euler's method and Heun's method is that in the previous method the slope at one point is considered but in Heun's method the average of two slopes are considered (as illustrated in Figure 6.3). Thus, the general form is

$$y_{i+1} = y_i + \frac{h}{2}\left[f\left(x_i, y_i\right) + f\left(x_i + h, y_i + hf\left(x_i, y_i\right)\right)\right],$$

(6.23)

where $x_{i+1} = x_i + h$, for $i = 0, 1, 2, \ldots$.

The Eq. (6.23) can be obtained by using the trapezoidal rule, that's why the local and global errors are of order h^3 and h^2.

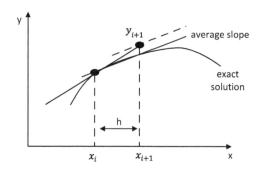

FIGURE 6.3   A graphical illustration of Heun's method.

Example 6.3: Use Heun's method to calculate the solution of $\frac{dy}{dx} = xy$ with $y(0) = 1$ for $x = 0, 0.2, 0.4, 0.6, 0.8$ and 1.

Solution: Here $h = 0.2, f\left(x, y\right) = xy, x_0 = 0, y_0 = 1,\ x_1 = 0.2, x_2 = 0.4, x_3 = 0.6, x_4 = 0.8$ and $x_5 = 1$.

By Heun's formula, for i = 0

$$y_1 = y_0 + \frac{h}{2}\left[f\left(x_0, y_0\right) + f\left(x_0 + h, y_0 + hf\left(x_0, y_0\right)\right)\right]$$

$$= 1 + \frac{0.2}{2}\left[0 + f\left(0.2, 1\right)\right]$$

$$= 1 + \frac{0.2}{2}\left[0 + 0.2\right] = 1.02,$$

for i = 1

$$y_2 = y_1 + \frac{h}{2}\left[f\left(x_1, y_1\right) + f\left(x_1 + h, y_1 + hf\left(x_1, y_1\right)\right)\right]$$

$$= 1.02 + \frac{0.2}{2}\left[\left(0.2\right)\left(1.02\right) + f\left(0.4, 1.02 + 0.2\left(0.2\right)\left(1.02\right)\right)\right]$$

$$= 1.02 + 0.1\left[\left(0.2040\right) + f\left(0.4, 1.0608\right)\right]$$

$$= 1.02 + 0.1\left[\left(0.2040\right) + 0.4243\right]$$

$$= 1.0828,$$

for i = 2, 3 and 4 the results will be $y_3 = 1.1963, y_4 = 1.3753,\ y_5 = 1.6448$.

Main steps for solving Heun's method in MATLAB: The procedure of Heun's method is similar to Euler's method in MATLAB; just that Euler's formula is replaced with Heun's formula, i.e.,

$y(i + 1) = y(i) + 0.5 * h * (func(x(i),y(i)) + func(x(i) + h,y(i) + h * func(x(i),y(i))))$.

```
% Mfile 1
function [z]=Heuns(func,y0,x0,h,xf)
% z is the output
% func, y0, x0, h and xf are the inputs
x=x0:h:xf; % Domain
y(1)=y0; % The initial value is used in the loop
for i=1:length(x)-1 % In order to make equal dimensions
  y(i+1)=y(i)+0.5*h*(func(x(i),y(i))+ ...
   func(x(i)+h,y(i)+h*func(x(i),y(i)))); % Heun's method
end
z=y;
```

```
% Mfile 2
function [z]=tys(x,y)
z=x*y;
end
```

```
>> % Command Window
>> Heuns(@tys,1,0,0.2,1)
```

6.1.2.1 Heun's Method for the Systems of Ordinary Differential Equations

Example 6.4: Solve $y'' + 3y' - 2y = 5$ with $y(0) = 0$, $y'(0) = 1$ and $h = 0.1$ and calculate $y(1)$.

Solution: Let

$$z = y',$$

so

$$z' = 5 - 3z + 2y.$$

With initial conditions reduced to

$$y(0) = 0, z(0) = 1.$$

We have $h = 0.1, x_0 = 0, y_0 = 0, z_0 = 1$.

Since $x_0 = 0, x_1 = x_0 + h = 0.1, x_2 = x_0 + 2h = 0.2$,

$$x_3 = x_0 + 3h = 0.3, x_4 = x_0 + 4h = 0.4,$$

$$x_5 = x_0 + 5h = 0.5, x_6 = x_0 + 6h = 0.6,$$

$$x_7 = x_0 + 7h = 0.7, x_8 = x_0 + 8h = 0.8,$$

$$x_9 = x_0 + 9h = 0.9, x_{10} = x_0 + 10h = 1.$$

By using Heun's formula for $i = 0$:

$$y_1 = y_0 + \frac{h}{2}\left[f(x_0,y_0,z_0) + f(x_0 + h, y_0 + hf(x_0,y_0,z_0), z_0 + hf(x_0,y_0,z_0))\right]$$

$$= 0 + \frac{0.1}{2}\left[1 + f(0.1, 0 + 0.1(1), 1 + 0.1(1))\right]$$

$$= \frac{0.1}{2}\left[1 + f(0.1, 0.1, 1.1)\right] = 0.1100,$$

$$z_1 = z_0 + \frac{h}{2}\left[g(x_0,y_0,z_0) + g(x_0 + h, y_0 + hg(x_0,y_0,z_0), z_0 + hg(x_0,y_0,z_0))\right]$$

$$= 1 + \frac{0.1}{2}\left[(5 - 3z_0 + 2y_0) + g(x_0 + h, y_0 + h(5 - 3z_0 + 2y_0), z_0 + h(5 - 3z_0 + 2y_0))\right]$$

$$= 1 + 0.0500\left[(5 - 3(1) + 2(0)) + g(0.1, 0 + 0.1(5 - 3(1) + 2(0)), 1 + 0.1(5 - 3(1) + 2(0)))\right]$$

$$= 1 + 0.0500\left[2 + g(0.1, 0.2, 1.2)\right]$$

$$= 1 + 0.0500\left[2 + 5 - 3(1.2) + 2(0.2)\right] = 1.1800.$$

By using the same procedure, the results for $i = 1,2,3...9$ become

$$\binom{y_2}{z_2} = \binom{0.2364}{1.3346}, \binom{y_3}{z_3} = \binom{0.3772}{1.4728}, \binom{y_4}{z_4} = \binom{0.5312}{1.6011}, \binom{y_5}{z_5} = \binom{0.6976}{1.7241},$$

$$\binom{y_6}{z_6} = \binom{0.8761}{1.8453}, \binom{y_7}{z_7} = \binom{1.0667}{1.9671}, \binom{y_8}{z_8} = \binom{1.2696}{2.0915}, \binom{y_9}{z_9} = \binom{1.4851}{2.2199}, \binom{y_{10}}{z_{10}} = \binom{1.7136}{2.3535}.$$

Main steps for solving Heun's method for the systems in MATLAB: The method is solved in the same way as Euler's method for the system of equations, just the formula is changed, i.e.,

y(:,i + 1) = y(:,i) + 0.5 * h * (func(x(i),y(:,i)) + func(x(i) ...
+ h,y(:,i) + h * func(x(i),y(:,i)))).

```
% Mfile 1
function [z]=Heuns_Method_System(func,y1,x0,h,xf)
% z is the output
% func, y1, x0,h and xf are the inputs
x=x0:h:xf;
y(:,1)=y1; % One column
for i=1:length(x)-1
    y(:,i+1)=y(:,i)+0.5*h*(func(x(i),y(:,i))+func(x(i)...
    +h,y(:,i)+h*func(x(i),y(:,i)))); % Improved Euler's formula
end
z=y'; % Column vectors
plot(x,z(:,2))
```

```
% Mfile 2
function [z]=tys1(x,y)
z=[y(2); 5-3*y(2) + 2*y(1)];
end
```

```
>> % Command Window
>> [z]= Heuns_Method_System(@tys1,[0 1],0,0.1,1)
```

6.1.3 Modified Euler's Method

This method uses Euler's method in which the slope used is taken at the midpoint of the interval (as shown in Figure 6.4), i.e.,

$$y_{i+1/2} = y_i + \frac{1}{2} hf\left(x_i, y_i\right). \tag{6.24}$$

This value is used to calculate the slope at the midpoint, i.e.,

$$y'_{i+1/2} = f\left(x_{i+1/2}, y_{i+1/2}\right). \tag{6.25}$$

Eq. (6.25) is the approximate average slope for the entire interval. This slope is used to extrapolate linearly from x_i to x_{i+1}

$$y_{i+1} = y_i + hf\left(x_{i+1/2}, y_{i+1/2}\right), \tag{6.26}$$

or

$$y_{i+1} = y_i + h\left[f\left(x_i + \frac{1}{2} h\ , y_i + \frac{1}{2} hf\left(x_i, y_i\right)\right)\right], \tag{6.27}$$

where $x_{i+1} = x_i + h,$ for $i = 1, 2, 3\ldots.$

This method can be obtained from trapezoidal rule, so its local and global errors are of order h^3 and h^2.

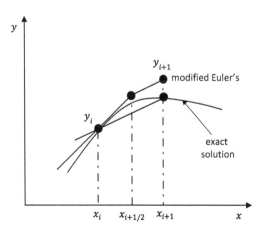

FIGURE 6.4 A graphical illustration of modified Euler's method.

Example 6.5: Use the modified Euler's method to calculate the numerical solution of the problem $\dfrac{dy}{dx} = xy$, $y(0) = 1$ for $x = 0, 0.2, 0.4, 0.6, 0.8$ and 1.

Solution: Here $h = 0.2, f(x,y) = xy, x_0 = 0, y_0 = 1,\ x_1 = 0.2, x_2 = 0.4, x_3 = 0.6, x_4 = 0.8$ and $x_5 = 1$.
By the modified Euler's formula, for i = 0

$$y_1 = y_0 + h\left[f\left(x_0 + \frac{1}{2}h, y_0 + \frac{1}{2}hf(x_0, y_0)\right)\right]$$

$$= 1 + 0.2\left[f\left(0 + \frac{1}{2}0.2, 1 + \frac{1}{2}0.2(0)\right)\right]$$

$$= 1 + 0.2\left[f(0.1, 1)\right] = 1 + 0.2[0.1] = 1.0200,$$

for i = 1

$$y_2 = y_1 + h\left[f\left(x_1 + \frac{1}{2}h, y_1 + \frac{1}{2}hf(x_1, y_1)\right)\right]$$

$$= 1.0200 + 0.2\left[f\left(0.2 + \frac{1}{2}0.2, 1.0200 + \frac{1}{2}0.2(0.2040)\right)\right]$$

$$= 1.0200 + 0.2\left[f(0.3, 1.0404)\right]$$

$$= 1.0200 + 0.2[0.3121] = 1.0824.$$

The remaining values are calculated by using the same procedure which gives
$y_3 = 1.1965$, $y_4 = 1.3723$ and $y_5 = 1.6391$.

Main steps for solving the modified Euler's method in MATLAB: The method is solved in the same way as Euler's method, just the formula is changed, i.e.,

y(i + 1) = y(i) + h * func(x(i) + 0.5 * h,y(i) + 0.5 * h * func(x(i),y(i))).

```
% Mfile 1
function [z]=modified_euler_Method(func,y0,x0,h,xf)
% z is the output
% func, y0, x0,h and xf are the inputs
x=x0:h:xf; % Doamin
y(1)=y0; % The initial value is used in the loop
for i=1:length(x)-1 % In order to make equal dimensions
    % Modified Euler's formula
    y(i+1)=y(i)+h*func(x(i)+0.5*h,y(i)+0.5*h*func(x(i),y(i)));
end
z=y;
```

```
% Mfile 2
function [z]=tys(x,y)
z=x*y;
end
```

```
>> % Command Window
>> modified_euler_Method(@tys,1,0,0.2,1)
```

6.1.3.1 Modified Euler's Method for the Systems of Ordinary Differential Equations

Example 6.6: Solve $y'' + 3y' - 2y = 5$ with $y(0) = 0$, $y'(0) = 1$ and $h = 0.1$, calculate $y(1)$.

Solution: Let

$$z = y',$$

so

$$z' = 5 - 3z + 2y.$$

With initial conditions

$$y(0) = 0, z(0) = 1.$$

We have $h = 0.1 x_0 = 0, y_0 = 0, z_0 = 1$.

Since $x_0 = 0$, $x_1 = x_0 + h = 0.1$, $x_2 = x_0 + 2h = 0.2$, $x_3 = x_0 + 3h = 0.3$, $x_4 = x_0 + 4h$

$= 0.4$, $x_5 = x_0 + 5h = 0.5$, $x_6 = x_0 + 6h = 0.6$, $x_7 = x_0 + 7h = 0.7$, $x_8 = x_0 + 8h$

$= 0.8$, $x_9 = x_0 + 9h = 0.9$, $x_{10} = x_0 + 10h = 1$

Using modified Euler's formula for $i = 0$

$$y_1 = y_0 + h \left[f\left(x_0 + \frac{1}{2}h, y_0 + \frac{1}{2}hf(x_0, y_0, z_0), z_0 + \frac{1}{2}hf(x_0, y_0, z_0) \right) \right]$$

$$= 0 + 0.1 \left[f\left(x_0 + \frac{1}{2}0.1, 0 + \frac{1}{2}0.1(1), 1 + \frac{1}{2}0.1(1) \right) \right]$$

$$= 0.1 \left[f\left(0 + \frac{1}{2}0.1, 0 + \frac{1}{2}0.1(1), 1 + \frac{1}{2}0.1(1) \right) \right]$$

$$= 0.1 \left[f(0.05, 0.05, 1.05) \right] 0.1$$

$$= 0.1(1.05) = 0.1050.$$

$$z_1 = z_0 + h \left[g\left(x_0 + \frac{1}{2}h, y_0 + \frac{1}{2}hg(x_0, y_0, z_0), z_0 + \frac{1}{2}hg(x_0, y_0, z_0) \right) \right]$$

$$= 1 + 0.1 \left[g\left(x_0 + \frac{1}{2}0.1, 0 + \frac{1}{2}0.1(5 - 3z_0 + 2y_0), 1 + \frac{1}{2}0.1(5 - 3z_0 + 2y_0) \right) \right]$$

$$= 1 + 0.1 \left[g\left(0 + \frac{1}{2}0.1, 0 + \frac{1}{2}0.1(5 - 3(1) + 2(0)), 1 + \frac{1}{2}0.1(5 - 3(1) + 2(0)) \right) \right]$$

$$= 1 + 0.1 \left[g(0.05, 0.1, 1.1) \right] = 1 + 0.1(5 - 3(1.1) + 2(0.1)) = 1.1900.$$

By using the same procedure, the results for $i = 1, 2, 3 \ldots 9$ become

$$\begin{pmatrix} y_2 \\ z_2 \end{pmatrix} = \begin{pmatrix} 0.2364 \\ 1.3346 \end{pmatrix}, \begin{pmatrix} y_3 \\ z_3 \end{pmatrix} = \begin{pmatrix} 0.3772 \\ 1.4728 \end{pmatrix}, \begin{pmatrix} y_4 \\ z_4 \end{pmatrix} = \begin{pmatrix} 0.5312 \\ 1.6011 \end{pmatrix}, \begin{pmatrix} y_5 \\ z_5 \end{pmatrix} = \begin{pmatrix} 0.6976 \\ 1.7241 \end{pmatrix},$$

$$\begin{pmatrix} y_6 \\ z_6 \end{pmatrix} = \begin{pmatrix} 0.8761 \\ 1.8453 \end{pmatrix}, \begin{pmatrix} y_7 \\ z_7 \end{pmatrix} = \begin{pmatrix} 1.0667 \\ 1.9671 \end{pmatrix}, \begin{pmatrix} y_8 \\ z_8 \end{pmatrix} = \begin{pmatrix} 1.2696 \\ 2.0915 \end{pmatrix}, \begin{pmatrix} y_9 \\ z_9 \end{pmatrix} = \begin{pmatrix} 1.4851 \\ 2.2199 \end{pmatrix}, \begin{pmatrix} y_{10} \\ z_{10} \end{pmatrix} = \begin{pmatrix} 1.7136 \\ 2.3535 \end{pmatrix}.$$

Main steps for solving the modified Euler's method for the systems in MATLAB: The method is solved in the same way as Euler's method for the system of equations, just the formula is changed, i.e.,

$y(:,i + 1) = y(:,i) + h * func(x(i) + 0.5 * h,y(:,i) + 0.5 * h * func(x(i),y(:,i)))$.

```
% Mfile 1
function [z]=modified_euler_Method_System(func,y1,x0,h,xf)
% z is the output
% func, y1, x0,h and xf are the inputs
x=x0:h:xf; % Domain
y(:,1)=y1; % One column
for i=1:length(x)-1
    % Modified Euler's formula
    y(:,i+1)=y(:,i)+h*func(x(i)+0.5*h,y(:,i)...
      +0.5*h*func(x(i),y(:,i)));
end
z=y'; % Column vectors
```

```
% Mfile 2
function [z]=tys1(x,y)
z=[y(2); 5-3*y(2)+2*y(1)];
end
```

```
>> % Command Window
>> [z]=modified_euler_Method_System(@tys1,[0 1],0,0.1,1)
```

6.1.4 RUNGE-KUTTA METHODS

The Taylor series approach is used to find different forms of Runge-Kutta methods without involving any derivative. The general form is:

$$y_{i+1} = y_i + \varnothing, \tag{6.28}$$

where $\varnothing$ denotes an increment function, which shows the slope over the interval. The general form of the increment function can be written as:

$$\varnothing = \alpha_1 k_1 + \alpha_2 k_2 + ... + \alpha_n k_n. \tag{6.29}$$

Here the α's represents the weighted averages and k's are the slopes

$$k_1 = hf(x_i, y_i),$$

$$k_2 = hf(x_i + hm_1, y_i + l_{11}k_1),$$

$$k_3 = hf(x_i + hm_2, y_i + l_{21}k_1 + l_{22}k_2),$$

$$\cdot$$
$$\cdot$$
$$\cdot$$

$$k_n = hf(x_i + hm_{n-1}, y_i + l_{n-1,1}k_1 + l_{n-1,2}k_2 ... l_{n-1,n-1}k_{n-1}). \tag{6.30}$$

By specifying different values of n along with several values of m's and l's various types of Runge-Kutta methods are generated.

6.1.4.1 Second Order Runge-Kutta Methods

The second order form of the above equation is:

$$y_{i+1} = y_i + \left(\alpha_1 k_1 + \alpha_2 k_2 \right), \tag{6.31}$$

where

$$k_1 = hf\left(x_i, y_i \right),$$

$$k_2 = hf\left(x_i + hm_1, y_i + l_{11}k_1 \right). \tag{6.32}$$

Here α_1, α_2, m_1 and l_{11} are calculated by setting the above equation equal to the Taylor series of second order. By doing this, three equations are achieved along with four constants. The equations are:

$$\alpha_1 + \alpha_2 = 1,$$

$$\alpha_2 m_1 = \frac{1}{2},$$

$$\alpha_2 l_{11} = \frac{1}{2}. \tag{6.33}$$

Here it's not possible to calculate these values unless one value is known. Suppose we are given the value of α_2. Then the equations become

$$\alpha_1 = 1 - \alpha_2,$$

$$l_{11} = m_1 = \frac{1}{2\alpha_2}. \tag{6.34}$$

Here we can choose infinite values of α_2 so there are infinite numbers of second order Runge-Kutta methods. Each method will give the same result.

The most used three methods are:

Heun Method $(\alpha_2 = 1/2)$: If we choose $\alpha_2 = 1/2$, then after putting the values of l_{11}, m_1 and α_1 we get

$$y_{i+1} = y_i + \left(\frac{1}{2} k_1 + \frac{1}{2} k_2 \right), \tag{6.35}$$

where

$$k_1 = hf\left(x_i, y_i \right),$$

$$k_2 = hf\left(x_i + h, y_i + k_1 \right). \tag{6.36}$$

This is Heun's technique.

Modified Euler's Method $(\alpha_2 = 1)$: If we choose $\alpha_2 = 1$, then after putting the values of l_{11}, m_1 and α_1 we have

$$y_{i+1} = y_i + k_2, \tag{6.37}$$

where

$$k_1 = hf\left(x_i, y_i\right),$$

$$k_2 = hf\left(x_i + h/2, y_i + k_1/2\right). \tag{6.38}$$

This is the modified Euler's method.

Ralston's Method ($\alpha_2 = 2/3$): If we choose $\alpha_2 = 2/3$, then after putting the values of l_{11}, m_1 and α_1 we have

$$y_{i+1} = y_i + \left(\frac{1}{3}k_1 + \frac{2}{3}k_2\right), \tag{6.39}$$

where

$$k_1 = hf\left(x_i, y_i\right),$$

$$k_2 = hf\left(x_i + \frac{3}{4}h, y_i + \frac{3}{4}k_1\right). \tag{6.40}$$

This is the Ralston's Method.

6.1.4.2 Classical Fourth Order Runge-Kutta Method

One of the most popular methods is the fourth order Runge-Kutta method. Like second order Runge-Kutta methods, it has an infinite number of versions. The commonly used version of fourth order Runge-Kutta method is

$$y_{i+1} = y_i + \frac{1}{6}\left(k_1 + 2k_2 + 2k_3 + k_4\right), \tag{6.41}$$

where

$$k_1 = hf\left(x_i, y_i\right),$$

$$k_2 = hf\left(x_i + \frac{h}{2}, y_i + \frac{k_1}{2}\right),$$

$$k_3 = hf\left(x_i + \frac{h}{2}, y_i + \frac{k_2}{2}\right),$$

$$k_4 = hf\left(x_i + h, y_i + k_3\right). \tag{6.42}$$

The fourth order Runge-Kutta method is like Heun's approach, the difference is that multiple estimates of slopes are developed which results in an improved average slope for the interval (as shown in Figure 6.5). The local truncation error is of order h^5. We can calculate the local truncation error of fourth order Runge-Kutta method when we approximate the accuracies equal to Taylor's series of order 4 without calculating the analytical differentiation $f(x, y)$. If the local truncation error is of order h^5 then the global truncation error is of order h^4.

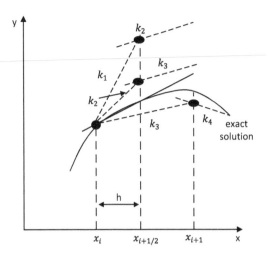

FIGURE 6.5 A graphical illustration of fourth order Runge-Kutta method.

Example 6.7: Use the fourth order Runge-Kutta method to calculate the numerical solution of the problem $\dfrac{dy}{dx} = xy$, $y(0) = 1$ for $x = 0, 0.2, 0.4, 0.6, 0.8$ and 1.

Solution: Here $h = 0.2, f(x,y) = xy, x_0 = 0, y_0 = 1,$ $x_1 = 0.2, x_2 = 0.4, x_3 = 0.6, x_4 = 0.8$ and $x_5 = 1$.
For $i = 0$

$$k_1 = hf(x_0, y_0) = hx_0y_0 = (0.2)(0)(1) = 0,$$

$$k_2 = hf\left(x_0 + \frac{h}{2}, y_0 + \frac{k_1}{2}\right) = h\left(x_0 + \frac{h}{2}\right)\left(y_0 + \frac{k_1}{2}\right)$$

$$= (0.2)\left(0 + \frac{0.2}{2}\right)(1+0) = 0.0200,$$

$$k_3 = hf\left(x_0 + \frac{h}{2}, y_0 + \frac{k_2}{2}\right) = (0.2)\left(0 + \frac{0.2}{2}\right)\left(1 + \frac{0.0200}{2}\right) = 0.0202,.$$

$$k_4 = hf(x_0 + h, y_0 + k_3) = (0.2)(0 + 0.2)(1 + 0.0202) = 0.0408,$$

$$y_1 = y_0 + \frac{1}{6}(k_1 + 2k_2 + 2k_3 + k_4)$$

$$= 1 + \frac{1}{6}\left(0 + 2(0.0200) + 2(0.0202) + 0.0408\right)$$

$$= 1.0202.$$

In a similar manner the remaining values for $i = 1, 2, 3$ and 4 are
$y_2 = 1.0833$, $y_3 = 1.1972$, $y_4 = 1.3771$, $y_5 = 1.6487$.

Main steps for solving the fourth order Runge-Kutta method in MATLAB: The output will be z. The function is 'func', y0, x0 are the initial values, h the increment and the final value will be xf.

1. The domain is defined as
   x = x0:h:xf.
2. The initial value y0 is defined as y(1) = y0.
3. The loop starts from 1 and ends at length(x) – 1.

4. The fourth order values of k_1, k_2, k_3 and k_4 are written.
5. Afterwards y_{i+1} is defined.
6. The second script is used to define the problem.
7. The command window will give output.

```
% Mfile 1
function [z]=fourth_order_RK(func,y0,x0,h,xf)
% z is the output
% func, y0, x0, h and xf are the inputs
x=x0:h:xf; % Domain
y(1)=y0; % The initial value is used in the loop
for i=1:length(x)-1 % In order to make equal dimensions
    k1=h*func(x(i),y(i));
     k2=h*func(x(i)+0.5*h,y(i)+0.5*k1);
     k3=h*func(x(i)+0.5*h,y(i)+0.5*k2);
     k4=h*func(x(i)+h,y(i)+k3);
     dely=(1/6)*(k1+2*k2+2*k3+k4);
     y(i+1)=y(i)+dely;
end
z=y;
```

```
% Mfile 2
function [z]=tys(x,y)
z=x*y;
end
```

```
>> % Command Window
>> [z]=fourth_order_RK(@tys,1,0,0.2,1)
```

6.1.4.3 Fourth Order Runge-Kutta Method for the Systems of Ordinary Differential Equations

Example 6.8: Solve $y'' + 3y' - 2y = 5$ with $y(0) = 0, y'(0) = 1$ and h = 0.25, calculate $y(1)$.

Solution: Let

$$z = y',$$

so

$$z' = 5 - 3z + 2y.$$

With initial conditions

$$y(0) = 0, z(0) = 1.$$

We have $h = 0.25, x_0 = 0, y_0 = 0, z_0 = 1.$

$$x_0 = 0, x_1 = x_0 + h = 0.2500,$$

Since $x_2 = x_0 + 2h = 0.5000,$

$$x_3 = x_0 + 3h = 0.7500, x_4 = x_0 + 4h = 1.0000.$$

For i = 0

$$k_1 = hf\left(x_0, y_0, z_0\right) = hz_0 = (0.25)(1) = 0.2500,$$

$$l_1 = hg\left(x_0, y_0, z_0\right) = h(5 - 3z_0 + 2y_0)$$

$$= (0.25)\left(5 - 3(1) + 2(0)\right)$$

$$= 0.5000,$$

$$k_2 = hf\left(x_0 + 0.5h, y_0 + 0.5k_1, z_0 + 0.5l_1\right)$$

$$= h(z_0 + 0.5l_1)$$

$$= (0.25)\left(1 + 0.5(0.5000)\right) = 0.3125,$$

$$l_2 = hg\left(x_0 + 0.5h, y_0 + 0.5k_1, z_0 + 0.5l_1\right)$$

$$= h\left(5 - 3(z_0 + 0.5l_1) + 2(y_0 + 0.5k_1)\right)$$

$$= 0.25\left(5 - 3(1 + 0.5(0.500)) + 2(0 + 0.5(0.2500))\right)$$

$$= 0.3750,$$

$$k_3 = hf\left(x_0 + 0.5h, y_0 + 0.5k_2, z_0 + 0.5l_2\right)$$

$$= h(z_0 + 0.5l_2)$$

$$= 0.25\left(1 + 0.5(0.3750)\right)$$

$$= 0.2969,$$

$$l_3 = hg\left(x_0 + 0.5h, y_0 + 0.5k_2, z_0 + 0.5l_2\right)$$

$$= h\left(5 - 3(z_0 + 0.5l_2) + 2(y_0 + 0.5k_2)\right)$$

$$= 0.25\left(5 - 3(1 + 0.5(0.3750)) + 2(0 + 0.5(0.3125))\right)$$

$$= 0.4375,$$

$$k_4 = hf\left(x_0 + h, y_0 + k_3, z_0 + l_3\right)$$

$$= h(z_0 + l_3)$$

$$= 0.25(1 + 0.4375)$$

$$= 0.3594,$$

$$l_4 = hg\left(x_0 + h, y_0 + k_3, z_0 + l_3\right)$$

$$= h\left(5 - 3(z_0 + l_3) + 2(y_0 + k_3)\right)$$

$$= 0.25\left(5 - 3(1 + 0.4375) + 2(0 + 0.2969)\right)'$$

$$= 0.3203$$

$$y_1 = y_0 + \frac{1}{6}\left(k_1 + 2k_2 + 2k_3 + k_4\right)$$

$$= 0 + \frac{1}{6}\left(0.2500 + 2(0.3125) + 2(0.2969) + 0.3594\right),$$

$$= 0.3047$$

$$z_1 = z_0 + \frac{1}{6}\left(l_1 + 2l_2 + 2l_3 + l_4\right)$$

$$= 1 + \frac{1}{6}\left(0.5000 + 2(0.3750) + 2(0.4375) + 0.3203\right)$$

$$= 1.4076.$$

By using the same procedure for i = 1, 2 and 3 the values are

$$\begin{pmatrix} y_2 \\ z_2 \end{pmatrix} = \begin{pmatrix} 0.6975 \\ 1.7262 \end{pmatrix}, \begin{pmatrix} y_3 \\ z_3 \end{pmatrix} = \begin{pmatrix} 1.1671 \\ 2.0305 \end{pmatrix} \text{ and } \begin{pmatrix} y_4 \\ z_4 \end{pmatrix} = \begin{pmatrix} 1.7146 \\ 2.3548 \end{pmatrix}.$$

Main steps for solving the fourth order Runge-Kutta method for the systems in MATLAB:
The output will be z and inputs are 'func', y1, x0, h and xf.

1. Firstly, the domain is defined
   x = x0:h:xf.
2. The given initial values $y_0 = 0$ and $z_0 = 1$ are defined as
   y(:,1) = y1.
3. The loop starts from
   i = 1:length(x) – 1 (domain and range must be equal if we want to plot).
4. For the fourth order Runge-Kutta method k_1, k_2, k_3 and k_4 are defined first then y is written so that y will pick the values of k_1, k_2, k_3 and k_4, i.e.,
   k1 = h * func(x(i),y(:,i)),
   k2 = h * func(x(i) + 0.5 * h,y(:,i) + 0.5 * k1),
   k3 = h * func(x(i) + 0.5 * h,y(:,i) + 0.5 * k2),
   k4 = h * func(x(i) + h,y(:,i) + k3),
   dely = (1/6) * (k1 + 2 * k2 + 2 * k3 + k4),
   y(:,i + 1) = y(:,i) + dely.
5. End the loop.
6. The second script is used to write the systems.
7. In the command window the results will be evaluated by using both scripts.

```
% Mfile 1
function [z]=fourth_order_RK_system(func,y0,x0,h,xf)
% z is the output
% func, y0, x0,h and xf are the inputs
x=x0:h:xf; % Domain
y(:,1)=y0; % The initial value is used in the loop
for i=1:length(x)-1 % In order to make equal dimensions
    k1=h*func(x(i),y(:,i));
    k2=h*func(x(i)+0.5*h,y(:,i)+0.5*k1);
    k3=h*func(x(i)+0.5*h,y(:,i)+0.5*k2);
    k4=h*func(x(i)+h,y(:,i)+k3);
    dely=(1/6)*(k1+2*k2+2*k3+k4);
    y(:,i+1)=y(:,i)+dely; % Rk4 formula
end
z=y'; % Column vectors
```

```
% Mfile 2
function [z]=tys1(x,y)
z=[y(2); 5-3*y(2)+2*y(1)];
end
```

```
>> % Command Window
>> [z]=fourth_order_RK_system(@tys1,[0 1],0,0.25,1)
```

6.1.4.4   Runge-Kutta-Fehlberg Technique

The Runge-Kutta-Fehlberg technique is used to control the error appearing while calculating the solution. It has the procedure to calculate whether the step size is appropriate or not. At each step, two different approximations for the solutions are made and compared. If the two approximations are approximately equal, then the approximation is acceptable. If the approximations are not close to each other than the step size is reduced. If the approximations agree to more significant digits than required, then the step size is increased. A better solution is calculated by using the Runge-Kutta method of order 5, i.e.,

$$\breve{y}_{i+1} = y_i + \frac{16}{135}k_1 + \frac{6656}{12825}k_3 + \frac{28561}{56430}k_4 - \frac{9}{50}k_5 + \frac{2}{55}k_6. \tag{6.43}$$

The local error in a Runge-Kutta method of order 4 is

$$y_{i+1} = y_i + \frac{25}{216}k_1 + \frac{1408}{2565}k_3 + \frac{2197}{4104}k_4 - \frac{1}{5}k_5, \tag{6.44}$$

where the coefficients are

$$k_1 = hf(x_i, y_i),$$

$$k_2 = hf\left(x_i + \frac{h}{4}, y_i + \frac{k_1}{4}\right),$$

$$k_3 = hf\left(x_i + \frac{3h}{8}, y_i + \frac{3k_1}{32} + \frac{9k_2}{32}\right),$$

$$k_4 = hf\left(x_i + \frac{12h}{13}, y_i + \frac{1932k_1}{2197} - \frac{7200k_2}{2197} + \frac{7296k_3}{2197}\right),$$

$$k_5 = hf\left(x_i + \frac{h}{2}, y_i + \frac{439k_1}{216} - 8k_2 + \frac{3680k_3}{513} - \frac{845k_4}{4104}\right),$$

$$k_6 = hf\left(x_i + \frac{h}{2}, y_i - \frac{8k_1}{27} + 2k_2 - \frac{3544k_3}{2565} + \frac{1859k_4}{4104} - \frac{11k_5}{40}\right). \tag{6.45}$$

The value of h is assumed randomly and then the coefficients are calculated and the values are substituted into $\breve{y}_{i+1}$ and y_{i+1}. After this the value of h is refined by the relation

$$R = \frac{\left|\breve{y}_{i+1} - y_{i+1}\right|}{h},$$

$$q = 0.84\left(\frac{\epsilon h}{R}\right)^{1/4}. \tag{6.46}$$

For $q < 1$ the chosen value h is rejected, and the calculations are repeated by using $h = qh$.

For $q \geq 1$, the computed value of h is accepted for the current iteration and may change in the next iteration. We will see a large number of modifications in the step size h. The derivation of Eq. (6.46) is available in advanced numerical books [5–67].

Example 6.9: Use the fifth order Runge-Kutta method to calculate the numerical solution of the problem $\dfrac{dy}{dx} = xy$, $y(0) = 1$ for $x = 0, 0.2, 0.4, 0.6, 0.8$ and 1 with tolerance $\varepsilon = 0.00001$.

Solution: Here $h = 0.2, f(x,y) = xy, x_0 = 0, y_0 = 1$, $x_1 = 0.2, x_2 = 0.4, x_3 = 0.6, x_4 = 0.8$ and $x_5 = 1$.
For $i = 0$

$$k_1 = hf(x_0, y_0) = hx_0 y_0 = (0.2)(0)(1) = 0,$$

$$k_2 = hf\left(x_0 + \frac{h}{4}, y_0 + \frac{k_1}{4}\right) = \left(x_0 + \frac{h}{4}\right)\left(y_0 + \frac{k_1}{4}\right)$$

$$= (0.2)\left(0 + \frac{0.2}{4}\right)\left(1 + \frac{0}{4}\right) = 0.0100,$$

$$k_3 = hf\left(x_0 + \frac{3h}{8}, y_0 + \frac{3k_1}{32} + \frac{9k_2}{32}\right)$$

$$= h(x_0 + \frac{3h}{8})\left(y_0 + \frac{3k_1}{32} + \frac{9k_2}{32}\right)$$

$$= 0.2\left(0 + \frac{3(0.2)}{8}\right)\left(1 + \frac{3(0)}{32} + \frac{9(0.0100)}{32}\right)$$

$$= 0.0150,$$

$$k_4 = hf\left(x_0 + \frac{12h}{13}, y_0 + \frac{1932k_1}{2197} - \frac{7200k_2}{2197} + \frac{7296k_3}{2197}\right)$$

$$= h(x_0 + \frac{12h}{13})\left(y_0 + \frac{1932k_1}{2197} - \frac{7200k_2}{2197} + \frac{7296k_3}{2197}\right)$$

$$= 0.2\left(0 + \frac{12(0.2)}{13}\right)\left(1 + \frac{1932(0)}{2197} - \frac{7200(0.0100)}{2197} + \frac{7296(0.0150)}{2197}\right)$$

$$= 0.0376,$$

$$k_5 = hf\left(x_0 + \frac{h}{2}, y_0 + \frac{439k_1}{216} - 8k_2 + \frac{3680k_3}{513} - \frac{845k_4}{4104}\right)$$

$$= h(x_0 + \frac{h}{2})\left(y_0 + \frac{439k_1}{216} - 8k_2 + \frac{3680k_3}{513} - \frac{845k_4}{4104}\right)$$

$$= 0.2\left(0 + \frac{0.2}{2}\right)\left(1 + \frac{439(0)}{216} - 8(0.0100) + \frac{3680(0.0150)}{513} - \frac{845(0.0376)}{4104}\right)$$

$$= 0.0408,$$

$$k_6 = hf\left(x_0 + \frac{h}{2}, y_0 - \frac{8k_1}{27} + 2k_2 - \frac{3544k_3}{2565} + \frac{1859k_4}{4104} - \frac{11k_5}{40}\right)$$

$$= h(x_0 + \frac{h}{2})\left(y_0 - \frac{8k_1}{27} + 2k_2 - \frac{3544k_3}{2565} + \frac{1859k_4}{4104} - \frac{11k_5}{40}\right)$$

$$= 0.2\left(0 + \frac{0.2}{2}\right)\left(1 - \frac{8(0)}{27} + 2(0.0100) - \frac{3544(0.0150)}{2565} + \frac{1859(0.0376)}{4104} - \frac{11(0.0408)}{40}\right)$$

$$= 0.0201.$$

Putting the values of coefficients gives

$$\overset{\varsigma}{y_1} = y_0 + \frac{16}{135}k_1 + \frac{6656}{12825}k_3 + \frac{28561}{56430}k_4 - \frac{9}{50}k_5 + \frac{2}{55}k_6$$

$$= 1 + \frac{16}{135}(0) + \frac{6656}{12825}(0.0150)$$

$$+ \frac{28561}{56430}(0.0376) - \frac{9}{50}(0.0408) + \frac{2}{55}(0.0201)$$

$$= 1.020201393136,$$

$$y_1 = y_0 + \frac{25}{216}k_1 + \frac{1408}{2565}k_3 + \frac{2197}{4104}k_4 - \frac{1}{5}k_5$$

$$= 1 + \frac{25}{216}(0) + \frac{1408}{2565}(0.0150)$$

$$+ \frac{2197}{4104}(0.0376) - \frac{1}{5}(0.0408)$$

$$= 1.020201384919,$$

and $s = 0.84 \left(\frac{0.00001(0.2)}{8.2163e-09} \right)^{1/4} = 3.3179 > 1.$

As $q \geq 1$, so the same value of h is used for the next iteration. The remaining values are mentioned in Table 6.1.

Main steps for solving the Runge-Kutta-Fehlberg technique in MATLAB: The output will be the solution which is y and the inputs will be the function 'func', the initial conditions y0 and x0 and the number of iterations n.

1. The increment is denoted by s.
2. The tolerance is represented by epsilon and has a small value 0.00001.
3. The initial values x0 and y0 are replaced with x(1) and y(1) because the loop starts from 1 so it is impossible for MATLAB to read x0 and y0.
   y(1) = y0,
   x(1) = x0.
4. Start the loop from 1 to n and after that write h = s because the increment may be modified depending on the conditions which will be elaborated.
5. The domain is defined by
   x(i + 1) = x(i) + h, where i = 1:n.

TABLE 6.1
Numerical Solutions of RKF-4 and RKF-5

	RKF-4	h		RKF-5
x_i	y_i		x_i	$\overset{\vee}{y}_i$
0	1.00000000	0.2	0	1.00000000
0.2	1.02020139	0.2	0.2	1.02020138
0.4	1.08328713	0.2	0.4	1.08328718
0.6	1.19721747	0.2	0.6	1.19721761
0.8	1.37712807	0.2	0.8	1.37712828
1	1.64872220	0.2	1	1.64872229

6. Next the values of k1, k2, k3, k4, k5 and k6 are written, k2 will pick the value from k1, k3 will pick value from k2 and so on. And these values are substituted in the formulas.

7. Now the increment remains the same or changed depending on the condition, i.e.,

R = abs(y(i + 1) − z(i + 1))/h,

q = 0.84 * ((h * epsilon)/R) ^ (1/4).

8. As in the given method, if q < 1 then we replace h by hq and if q > 1 then replace h by h. This can be sorted out by if and elseif conditions.

9. The next step is very important, which is to write s = h, and from this replacement the values of h are modified in the next step if q < 1 and remain the same if q > 1.

10. For the column vector write

y = y'.

11. The second script is used to write the function and both files are used to implement the results on the command window.

```
% Mfile 1
function [y]=fifth_order_RK(func,y0,x0,n)
% z is output
% func, y0, x0, h and n are the inputs
s=0.2; % Increment
epsilon=0.00001; % Tolerance
y(1)=y0; % The initial value y0
x(1)=x0; % The initial value x0
for i=1:n % In order to make equal dimensions
  h=s;
x(i+1)=x(i)+h;
 k1=h*func(x(i),y(i));
  k2=h*func(x(i)+(1/4)*h,y(i)+(1/4)*k1);
  k3=h*func(x(i)+(3/8)*h,y(i)+(3/32)*k1+(9/32)*k2);
  k4=h*func(x(i)+(12/13)*h,y(i)+(1932/2197)*k1-...
    (7200/2197)*k2+(7296/2197)*k3);
  k5=h*func(x(i)+h,y(i)+(439/216)*k1-(8)*k2+(3680/513)...
    *k3-(845/4104)*k4);
  k6=h*func(x(i)+0.5*h,y(i)-(8/27)*k1+(2)*k2-...
    (3544/2565)*k3+(1859/4104)*k4-(11/40)*k5);
  dely=(16/135)*k1+(6656/12825)*k3+(28561/56430)*...
    k4-(9/50)*k5+(2/55)*k6;
  delz=(25/216)*k1+(1408/2565)*k3+(2197/4104)*k4-(1/5)*k5;
z(i+1)=y(i)+delz;% Fourth order local error in Runge-Kutta
  y(i+1)=y(i)+dely; % Fifth order Runge-Kutta method
    % For the choice of h
  R=abs(y(i+1)-z(i+1))/h;
   q=0.84*((h*epsilon)/R)^(1/4);
 if q<1
 h=h*q;
  else q>1
  h=h;
end
s=h;
end
y=y';
```

```
% Mfile 2
function [z]=tys(x,y)
z=x*y;
end
```

```
>> % Command Window
>> [y]=fifth_order_RK(@tys,1,0,7)
```

6.1.4.5 Runge-Kutta-Fehlberg Technique for the Systems of Ordinary Differential Equations

Example 6.10: Solve $y'' + 3y' - 2y = 5$ with $y(0) = 0, y'(0) = 1$ and h = 0.25, calculate $y(1) = ?$

Solution: Let

$$z = y',$$

so

$$z' = 5 - 3z + 2y,$$

with initial conditions

$$y(0) = 0, z(0) = 1.$$

We have $h = 0.25, x_0 = 0, y_0 = 0, z_0 = 1$.
For $x_0 = 0, x_1 = x_0 + h = 0.2500$.
For i = 0

$$k_1 = hf(x_0, y_0, z_0) = hz_0 = (0.25)(1) = 0.2500,$$

$$l_1 = hg(x_0, y_0, z_0) = h(5 - 3z_0 + 2y_0)$$

$$= 0.25(5 - 3(1) + 2(0))$$

$$= 0.5000,$$

$$k_2 = hf\left(x_0 + \left(\frac{1}{4}\right)h, y_0 + \left(\frac{1}{4}\right)k_1, z_0 + \left(\frac{1}{4}\right)l_1\right)$$

$$= h(z_0 + \left(\frac{1}{4}\right)l_1)$$

$$= 0.25(1 + \left(\frac{1}{4}\right)(0.500)) = 0.2812,$$

$$l_2 = hg\left(x_0 + \left(\frac{1}{4}\right)h, y_0 + \left(\frac{1}{4}\right)k_1, z_0 + \left(\frac{1}{4}\right)l_1\right)$$

$$= h\left(5 - 3\left(z_0 + \left(\frac{1}{4}\right)l_1\right) + 2\left(y_0 + \left(\frac{1}{4}\right)k_1\right)\right)$$

$$= 0.25\left(5 - 3\left(1 + \left(\frac{1}{4}\right)(0.5000)\right) + 2\left(0 + \left(\frac{1}{4}\right)(0.25)\right)\right)$$

$$= 0.4375,$$

$$k_3 = hf\left(x_0 + \left(\frac{3}{8}\right)h, y_0 + \left(\frac{3}{32}\right)k_1 + \left(\frac{9}{32}\right)k_2, z_0 + \left(\frac{3}{32}\right)l_1 + \left(\frac{9}{32}\right)l_2\right)$$

$$= h\left(z_0 + \left(\frac{3}{32}\right)l_1 + \left(\frac{9}{32}\right)l_2\right)$$

$$= 0.25\left(1 + \left(\frac{3}{32}\right)(0.5000) + \left(\frac{9}{32}\right)(0.4375)\right)$$

$$= 0.2925,$$

$$l_3 = hg\left(x_0 + \left(\frac{3}{8}\right)h, y_0 + \left(\frac{3}{32}\right)k_1 + \left(\frac{9}{32}\right)k_2, z_0 + \left(\frac{3}{32}\right)l_1 + \left(\frac{9}{32}\right)l_2\right)$$

$$= h\left(5 - 3\left(z_0 + \left(\frac{3}{32}\right)l_1 + \left(\frac{9}{32}\right)l_2\right) + 2\left(y_0 + \left(\frac{3}{32}\right)k_1 + \left(\frac{9}{32}\right)k_2\right)\right)$$

$$= 0.25\left(5 - 3\left(1 + \left(\frac{3}{32}\right)(0.5000) + \left(\frac{9}{32}\right)(0.4375)\right)\right.$$

$$\left. + 2\left(0 + \left(\frac{3}{32}\right)(0.2500) + \left(\frac{9}{32}\right)(0.2812)\right)\right)$$

$$= 0.4238,$$

$$k_4 = hf\left(x_0 + \left(\frac{12}{13}\right)h, y_0 + \left(\frac{1932}{2197}\right)k_1 - \left(\frac{7200}{2197}\right)k_2\right.$$

$$\left. + \left(\frac{7296}{2197}\right)k_3, z_0 + \left(\frac{1932}{2197}\right)l_1 - \left(\frac{7200}{2197}\right)l_2 + \left(\frac{7296}{2197}\right)l_3\right)$$

$$= h\left(z_0 + \left(\frac{1932}{2197}\right)l_1 - \left(\frac{7200}{2197}\right)l_2 + \left(\frac{7296}{2197}\right)l_3\right)$$

$$= 0.25\left(1 + \left(\frac{1932}{2197}\right)(0.5000) - \left(\frac{7200}{2197}\right)(0.4375) + \left(\frac{7296}{2197}\right)(0.4238)\right)$$

$$= 0.3534,$$

$$l_4 = hg\left(x_0 + \left(\frac{12}{13}\right)h, y_0 + \left(\frac{1932}{2197}\right)k_1 - \left(\frac{7200}{2197}\right)k_2 + \left(\frac{7296}{2197}\right)k_3,\right.$$

$$z_0 + \left(\frac{1932}{2197}\right)l_1 - \left(\frac{7200}{2197}\right)l_2 + \left(\frac{7296}{2197}\right)l_3\right)$$

$$= h\left(5 - 3\left(z_0 + \left(\frac{1932}{2197}\right)l_1 - \left(\frac{7200}{2197}\right)l_2 + \left(\frac{7296}{2197}\right)l_3\right)\right.$$

$$\left. + 2\left(y_0 + \left(\frac{1932}{2197}\right)k_1 - \left(\frac{7200}{2197}\right)k_2 + \left(\frac{7296}{2197}\right)k_3\right)\right)$$

$$= 0.25 \left(5 - 3 \left(1 + \left(\frac{1932}{2197} \right) (0.5000) - \left(\frac{7200}{2197} \right) (0.4375) + \left(\frac{7296}{2197} \right) (0.4238) \right) \right.$$

$$\left. + 2 \left(0 + \left(\frac{1932}{2197} \right) (0.2500) - \left(\frac{7200}{2197} \right) (0.2812) + \left(\frac{7296}{2197} \right) (0.2925) \right) \right)$$

$$= 0.3247,$$

$$k_5 = hf \left(x_0 + h, y_0 + \left(\frac{439}{416} \right) k_1 - (8) k_2 + \left(\frac{3680}{513} \right) k_3 - \left(\frac{845}{4104} \right) k_4, \right.$$

$$z_0 + \left(\frac{3680}{513} \right) l_1 - \left(\frac{7200}{2197} \right) l_2 + \left(\frac{7296}{2197} \right) l_3 \right)$$

$$= h \left(z_0 + \left(\frac{3680}{513} \right) l_1 - \left(\frac{7200}{2197} \right) l_2 + \left(\frac{7296}{2197} \right) l_3 \right)$$

$$= 0.25 \left(1 + \left(\frac{3680}{513} \right) (0.5000) - \left(\frac{7200}{2197} \right) (0.4375) + \left(\frac{7296}{2197} \right) (0.4238) \right)$$

$$= 0.3724,$$

$$l_5 = hg \left(x_0 + h, y_0 + \left(\frac{439}{416} \right) k_1 - (8) k_2 + \left(\frac{3680}{513} \right) k_3 - \left(\frac{845}{4104} \right) k_4, \right.$$

$$z_0 + \left(\frac{439}{416} \right) l_1 - (8) l_2 + \left(\frac{3680}{513} \right) l_3 - \left(\frac{845}{4104} \right) l_4 \right)$$

$$= h (5 - 3 \left(z_0 + \left(\frac{439}{416} \right) l_1 - (8) l_2 + \left(\frac{3680}{513} \right) l_3 - \left(\frac{845}{4104} \right) l_4 \right)$$

$$+ 2 \left(y_0 + \left(\frac{439}{416} \right) k_1 - (8) k_2 + \left(\frac{3680}{513} \right) k_3 - \left(\frac{845}{4104} \right) k_4 \right)$$

$$= 0.25 (5 - 3 \left(1 + \left(\frac{439}{416} \right) (0.5000) - (8)(0.4375) + \right.$$

$$\left(\frac{3680}{513} \right) (0.4238) - \left(\frac{845}{4104} \right) (0.3247) \right) + 2 \left(0 + \left(\frac{439}{416} \right) (0.2500) \right.$$

$$- (8)(0.2812) + \left(\frac{3680}{513} \right) (0.2925) - \left(\frac{845}{4104} \right) (0.3534) \right)$$

$$= 0.2745,$$

$$k_6 = hf \left(x_0 + \left(\frac{1}{2} \right) h, y_0 - \left(\frac{8}{27} \right) k_1 + (2) k_2 - \left(\frac{3544}{2565} \right) k_3 + \left(\frac{1859}{4104} \right) k_4 - \left(\frac{11}{40} \right) k_5, \right.$$

$$z_0 - \left(\frac{8}{27} \right) l_1 + (2) l_2 - \left(\frac{3544}{2565} \right) l_3 + \left(\frac{1859}{4104} \right) l_4 - \left(\frac{11}{40} \right) l_5 \right)$$

$$= h\left(z_0 - \left(\frac{8}{27}\right) l_1 + (2) l_2 - \left(\frac{3544}{2565}\right) l_3 + \left(\frac{1859}{4104}\right) l_4 - \left(\frac{11}{40}\right) l_5 \right)$$

$$= 0.25\left(1 - \left(\frac{8}{27}\right)(0.5000) + (2)(0.4375) - \left(\frac{3544}{2565}\right)(0.4238) + \left(\frac{1859}{4104}\right)(0.3534) - \left(\frac{11}{40}\right)(0.2745) \right)$$

$$= 0.3032,$$

$$l_6 = hg\left(x_0 + \left(\frac{1}{2}\right)h, y_0 - \left(\frac{8}{27}\right)k_1 + (2)k_2 - \left(\frac{3544}{2565}\right)k_3 + \left(\frac{1859}{4104}\right)k_4 - \left(\frac{11}{40}\right)k_5, \right.$$

$$z_0 - \left(\frac{8}{27}\right)l_1 + (2)l_2 - \left(\frac{3544}{2565}\right)l_3 + \left(\frac{1859}{4104}\right)l_4 - \left(\frac{11}{40}\right)l_5 \right)$$

$$= h(5 - 3\left(z_0 - \left(\frac{8}{27}\right)l_1 + (2)l_2 - \left(\frac{3544}{2565}\right)l_3 + \left(\frac{1859}{4104}\right)l_4 - \left(\frac{11}{40}\right)l_5 \right)$$

$$+ 2\left(y_0 - \left(\frac{8}{27}\right)k_1 + (2)k_2 - \left(\frac{3544}{2565}\right)k_3 + \left(\frac{1859}{4104}\right)k_4 - \left(\frac{11}{40}\right)k_5 \right)$$

$$= 0.25(5 - 3\left(1 - \left(\frac{8}{27}\right)(0.5000) + (2)(0.4375) - \left(\frac{3544}{2565}\right)(0.4238) + \left(\frac{1859}{4104}\right)(0.3534) - \left(\frac{11}{40}\right)(0.2745) \right)$$

$$+ 2\left(0 - \left(\frac{8}{27}\right)(0.2500) + (2)(0.4375) - \left(\frac{3544}{2565}\right)(0.4238) + \left(\frac{1859}{4104}\right)(0.3534) - \left(\frac{11}{40}\right)0.3724 \right)$$

$$= 0.4113,$$

$$y_1 = y_0 + \left(\frac{16}{135}\right)k_1 + \left(\frac{6656}{12825}\right)k_3 + \left(\frac{28561}{56430}\right)k_4$$

$$- \left(\frac{9}{50}\right)k_5 + \left(\frac{2}{55}\right)k_6$$

$$= 0 + \left(\frac{16}{135}\right)(0.2500) + \left(\frac{6656}{12825}\right)(0.2925)$$

$$+ \left(\frac{28561}{56430}\right)(0.3534) - \left(\frac{9}{50}\right)(0.3724) + \left(\frac{2}{55}\right)(0.3032)$$

$$= 0.3042,$$

$$y'_1 = y'_0 + \left(\frac{25}{216}\right)k_1 + \left(\frac{1408}{2565}\right)k_3$$

$$+ \left(\frac{2197}{4104}\right)k_4 - \left(\frac{1}{5}\right)k_5$$

$$= y'_0 + \left(\frac{25}{216}\right)k_1 + \left(\frac{1408}{2565}\right)k_3 + \left(\frac{2197}{4104}\right)k_4 - \left(\frac{1}{5}\right)k_5$$

$$= 1.4094,$$

$$z_1 = z_0 + \left(\frac{16}{135}\right)k_1 + \left(\frac{6656}{12825}\right)k_3$$

$$+ \left(\frac{28561}{56430}\right)k_4 - \left(\frac{9}{50}\right)k_5$$

$$+ \left(\frac{2}{55}\right)k_6 = 0.3043,$$

$$z'_1 = z'_0 + \left(\frac{25}{216}\right)k_1 + \left(\frac{1408}{2565}\right)k_3 + \left(\frac{2197}{4104}\right)k_4 - \left(\frac{1}{5}\right)k_5$$

$$= 1.4091,$$

$$s = 0.84 \left(\frac{0.00001(0.25)}{3.7435e - 04}\right)^{1/4}$$

$$= 0.3396 < 1.$$

So h remains same for next iteration. Table 6.2 shows the remaining results of the problem.

Main steps for solving the Runge-Kutta-Fehlberg technique for the systems of ordinary differential equations in MATLAB: The output will be y, the inputs are the function 'func', the initial conditions y0 and x0 and the number of iterations n.

1. The increment is q.
2. The tolerance is 0.00001.
3. Replacing the initial values x0 and y0 by x(1) and y(:,1), y is in the form of a column vector, i.e.,
   y(:,1) = y0,
   x(1) = x0.
4. The loop starts from i = 1:n and after that write h = s as an increment which may be modified under the conditions which will be elaborated.
5. The domain is
   x(i + 1) = x(i) + h.
6. The values of k1, k2, k3, k4, k5, k6, y and z are implemented.
7. Now the increment remains the same or changed by the given condition.
   R = abs(y(1,i + 1) − z(1,i + 1))/h,
   s = 0.84 * ((epsilon)/R) ^ (1/4).
8. If s < 1 then we replace h by hs and if s > 1 then replace h by h. In MATLAB if and else condition is used.

TABLE 6.2
Numerical Solutions of RKF-4 and RKF-5

RKF-4			h	RKF-5		
x_i	y			x_i	z	
0	0	1.0000	0.25	0	0	1.0000
0.25	0.3043	1.4091	0.25	0.25	0.3043	1.4091
0.5	0.4287	1.5222	0.25	0.5	0.4287	1.5222
0.75	0.5626	1.6293	0.25	0.75	0.5626	1.6293
1	0.7053	1.7331	0.25	1	0.7053	1.7331

9. For modification in h, write q = h.
10. For the column vector write

   y = y'.

```
% Mfile 1
function [y]=fifth_order_Fehlberg_system(func,y0,x0,n)
% z is the output
% func, y0, x0, h and n are the inputs
q=0.25; % Increment
epsilon=0.00001; % Tolerance
y(:,1)=y0 % The initial value y0
x(1)=x0; % The initial value x0
for i=1:n
  h=q;
 x(i+1)=x(i)+h;
 k1=h*func(x(i),y(:,i));
 k2=h*func(x(i)+(1/4)*h,y(:,i)+(1/4)*k1);
 k3=h*func(x(i)+(3/8)*h,y(:,i)+(3/32)*k1+(9/32)*k2);
 k4=h*func(x(i)+(12/13)*h,y(:,i)+(1932/2197)*k1-...
  (7200/2197)*k2+(7296/2197)*k3);
 k5=h*func(x(i)+h,y(:,i)+(439/216)*k1-(8)*k2+(3680/513)...
  *k3-(845/4104)*k4);
 k6=h*func(x(i)+0.5*h,y(:,i)-(8/27)*k1+(2)*k2-...
  (3544/2565)*k3+(1859/4104)*k4-(11/40)*k5);
 dely=(16/135)*k1+(6656/12825)*k3+(28561/56430)*...
  k4-(9/50)*k5+(2/55)*k6;
 delz=(25/216)*k1+(1408/2565)*k3+(2197/4104)*k4-(1/5)*k5;
 z(:,i+1)=y(:,i)+delz;% Fourth order local error in Runge-Kutta
 y(:,i+1)=y(:,i)+dely; % Fifth order Runge-Kutta method
 % for the choice of h
 R=abs(y(1,i+1)-z(1,i+1))/h;
  s=0.84*((epsilon)/R)^(1/4);
 if s<1
 h=h*s;
 else s>1
 h=h;
 end
end
q=h;
end
y=y';
```

```
% Mfile 2
function [z]=tys1(x,y)
z=[y(2); 5-3*y(2)+2*y(1)];
end
```

```
>> % Command Window
>> fifth_order_Fehlberg_system(@tys1,[0 1],0,4)
```

6.1.5   MULTISTEP METHODS

In previous sections, the one-step methods are described and state that single mesh information can be used to compute the values of dependent variables (as shown in Figure 6.6(a)). However, in multistep methods, more than one previous mesh is used.

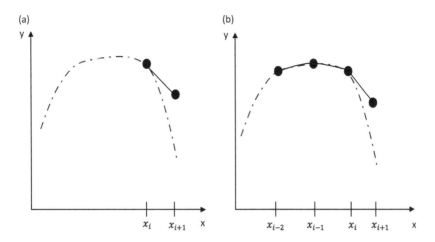

FIGURE 6.6 A graphical illustration of (a) single step and (b) multistep methods.

The dependent variable y's are said to be open formulas when y_{i+1} is written explicitly in terms of the known values of the function at x_{i-1}, x_i, etc. Similarly, the methods which include the unknown values of dependent variable y in the function are said to be closed formulas. The multistep methods are not self-started, since only one initial condition at $x = x_0$ is given, but we need more inputs to start the calculations. Therefore, one-step methods will be used to start the problem. When the required information is completed then we will use multistep methods.

6.1.5.1 Adams Multistep Methods

The Adams open or Adams-Bashforth formulas can be obtained by applying the backward finite difference approximation in the forward Taylor series expansion. The Taylor series expansion can be written as

$$y_{i+1} = y_i + hf(x_i) + \frac{h^2}{2}f'(x_i) + \frac{h^3}{6}f''(x_i) + \dots \tag{6.47}$$

By applying the backward finite difference approximation to $f'(x_i) = \dfrac{f(x_i) - f(x_{i-1})}{h}$ we get

$$y_{i+1} = y_i + hf(x_i) + \frac{h^2}{2}\frac{f(x_i) - f(x_{i-1})}{h} + O(h^3),$$

or

$$y_{i+1} = y_i + h\left(\frac{3}{2}f(x_i) - \frac{1}{2}f(x_{i-1})\right) + O(h^3). \tag{6.48}$$

The Eq. (6.48) is known as the second order open Adams formula. In a similar manner, the third order formula can be obtained by applying backward finite difference approximation of $f'(x_i)$ and $f''(x_i)$, i.e.,

$$f''(x_i) = \frac{f(x_i) - 2f(x_{i-1}) + f(x_{i-2})}{h^2} + O(h), \tag{6.49}$$

$$f'(x_i) = \frac{3f(x_i) - 4f(x_{i-1}) + f(x_{i-2})}{2h} + O(h^2). \tag{6.50}$$

Substituting Eqs. (6.49) and (6.50) in Eq. (6.47) we get

$$y_{i+1} = y_i + h\left(\frac{23}{12}f(x_i) - \frac{16}{12}f(x_{i-1}) + \frac{5}{12}f(x_{i-2})\right) + O(h^4). \tag{6.51}$$

Thus, the general form of the Adams-Bashforth method can be written as

$$y_{i+1} = y_i + h\sum_{k=1}^{n}\delta_{nk}f(x_{i+1-k}) + O(h^{n+1}). \tag{6.52}$$

The truncation error per step is $O(h^{n+1})$. The values of δ_{nk} for the Adams-Bashforth method is mentioned in Table 6.3.

The Adams closed or Adams-Moulton formulas can be obtained by applying the backward Taylor series expansion, i.e.,

$$y_i = y_{i+1} - hf(x_{i+1}) + \frac{h^2}{2}f'(x_{i+1}) - \frac{h^3}{6}f''(x_{i+1}) + \ldots \tag{6.53}$$

Or

$$y_{i+1} = y_i + hf(x_{i+1}) - \frac{h^2}{2}f'(x_{i+1}) + \frac{h^3}{6}f''(x_{i+1}) + \ldots \tag{6.54}$$

Thus, applying the backward finite difference approximation to $f'(x_{i+1}) = \dfrac{f(x_{i+1}) - f(x_i)}{h}$, we will get the second order closed formula, i.e.,

$$y_{i+1} = y_i + h\left(\frac{1}{2}f(x_i) + \frac{1}{2}f(x_{i+1})\right) + O(h^3). \tag{6.55}$$

Hence, by applying the same procedure we will get higher order closed formulas. The general form is written as

$$y_{i+1} = y_i + h\sum_{k=0}^{n-1}\delta^*_{nk}f(x_{i+1-k}) + O(h^{n+1}). \tag{6.56}$$

The values of δ^*_{nk} for the Adams closed or Adams-Moulton formulas are mentioned in Table 6.4.

TABLE 6.3

The Values of δ_{nk} for the Adams-Bashforth Method

n	δ	k=1	k=2	k=3	k=4	k=5	k=6
1	δ_{1k}	1					
2	$2\delta_{2k}$	3	−1				
3	$12\delta_{3k}$	23	−16	5			
4	$24\delta_{4k}$	55	−59	37	−9		
5	$720\delta_{5k}$	1901	−2774	2616	−1274	251	
6	$1440\delta_{6k}$	4277	−7923	9982	−7298	2877	−475

TABLE 6.4

The Values of δ^*_{nk} for the Adams closed or Adams-Moulton Formulas

n	δ^*	k=0	k=1	k=2	k=3	k=4	k=5
1	δ^*_{1k}	1					
2	$2\delta^*_{2k}$	1	1				
3	$12\delta^*_{3k}$	5	8	−1			
4	$24\delta^*_{4k}$	9	19	−5	1		
5	$720\delta^*_{5k}$	251	646	−264	106	−19	
6	$1440\delta^*_{6k}$	475	1427	−798	482	−173	27

6.1.5.2 Predictor-Corrector Methods by using Adams Formulas

The combination of an explicit method to predict and an implicit to modify the predication is called a predictor-corrector method. The Adams method can be used as a predictor-corrector (the Adams-Bashforth formulas as a predictor and the Adams-Moulton formulas as a corrector). If we choose the fourth order formula from Table 6.3, we will get the approximation $y_{n+1,r}$ as a predictor:

$$y_{i+1,r} = y_i + \left(\frac{h}{24}\right)\left(55y'_i - 59y'_{i-1} + 37y'_{i-2} - 9y'_{i-3}\right). \tag{6.57}$$

This approximation is modified by putting $y_{n+1,r}$ into a corrector (the values of the fourth order Adams-Moulton are mentioned in Table 6.4)

$$y_{i+1,c} = y_i + \left(\frac{h}{24}\right)\left[9y'_{i+1,r} + 19y'_i - 5y'_{i-1} + y'_{i-2}\right]. \tag{6.58}$$

In Eqs. (6.57) and (6.58), three values of y are calculated first. In this book we will use the fourth order Runge-Kutta method to calculate the starting values y_1, y_2 and y_3.

Example 6.11: Given $\frac{dy}{dx} = xy$, $y(0) = 1$, use fourth order Runge-Kutta method to calculate y(0.2), y(0.4) and y(0.6) and then use predictor-corrector method to calculate y(0.8) and y(1).

Solution: Here $h = 0.2, f(x,y) = xy, x_0 = 0, y_0 = 1,\ x_1 = 0.2, x_2 = 0.4, x_3 = 0.6, x_4 = 0.8$ and $x_5 = 1$. By using fourth order Runge-Kutta method.
    For i = 0

$$k_1 = hf(x_0, y_0) = x_0 y_0 = (0)(1) = 0,$$

$$k_2 = hf\left(x_0 + \frac{h}{2}, y_0 + \frac{k_1}{2}\right) = h\left(x_0 + \frac{h}{2}\right)\left(y_0 + \frac{k_1}{2}\right)$$

$$= (0.2)\left(0 + \frac{0.2}{2}\right)\left(1 + \frac{0}{2}\right) = 0.0200,$$

$$k_3 = hf\left(x_0 + \frac{h}{2}, y_0 + \frac{k_2}{2}\right) = h\left(x_0 + \frac{h}{2}\right)\left(y_0 + \frac{k_2}{2}\right)$$

$$= (0.2)\left(0 + \frac{0.2}{2}\right)\left(1 + \frac{0.0200}{2}\right) = 0.0202,$$

$$k_4 = hf\left(x_0 + h, y_0 + k_3\right) = h(x_0 + h)(y_0 + k_3)$$

$$= 0.2(0 + 0.2)(1 + 0.0202) = 0.0408,$$

$$y_1 = y_0 + \frac{1}{6}\left(k_1 + 2k_2 + 2k_3 + k_4\right)$$

$$= 1 + \frac{1}{6}\left(0 + 2(0.0200) + 2(0.0202) + 0.0408\right)$$

$$= 1.0202.$$

The remaining values of y for i = 1 and 2 are

$$y_2 = 1.0833, \ y_3 = 1.1972.$$

Now by using the fourth order Adam's formula
Predictor:

$$y_{4,r} = y_3 + \left(\frac{h}{24}\right)\left(55y'_3 - 59y'_2 + 37y'_1 - 9y'_0\right),$$

$$y_{4,r} = 1.1972 + \left(\frac{0.2}{24}\right)\left[55(0.7183) - 59(0.4333) + 37(0.2040) - 9(0)\right],$$

$$y_{4,r} = 1.3763.$$

Corrector:

$$y_{4,c} = y_3 + \left(\frac{h}{24}\right)\left[9y'_{4,r} + 19y'_3 - 5y'_2 + y'_1\right],$$

$$y_{4,c} = 1.1972 + \left(\frac{0.2}{24}\right)\left[9(1.1011) + 19(0.7183) - 5(0.4333) + (02040)\right],$$

$$y_{4,c} = 1.3772.$$

Similarly for i = 4 we have the results of the form

$$y_{5,r} = 1.6473, y_{5,c} = 1.6488.$$

Main steps for solving the predictor-corrector (the Adams-Bashforth formulas as a predictor and the Adams-Moulton formulas as a corrector) in MATLAB: The output is z, the inputs are the function 'func', the initial conditions y0 and x0, step size h and final point of the domain xf.

1. The domain is defined by
   x = x0:h:xf.
2. The initial value y0 is replaced with y(1) because in the next step the loop starts from 1 so it is impossible for MATLAB to read y0.
   y(1) = y0.
3. The loop starts from i = 1:3 and the fourth order Runge-Kutta method is used to calculate the first three values of y.
4. The remaining values are calculated by Adam's method and the loop starts from i = 4:length(x) − 1. And the formula for the predictor is written

s = y(i) + (h/24) * (55 * func(x(i),y(i)) – 59 * func(x(i – 1),y(i –1)) + 37 * func(x(i – 2),y(i – 2)) …
  – 9 * func(x(i – 3),y(i – 3))).
5. This value is used in the function as
   ss = func(x(i + 1),s).
6. This value ss is used in corrector formula as
   y(i + 1) = y(i) + (h/24) * (9 * ss + 19 * func(x(i),y(i))– …
   5 * func(x(i – 1),y(i – 1)) + func((i – 2),y(i – 2))).
7. At the end the function y is replaced with z.
8. The second script is used to write the function and both files are used to implement the results on the command window.

```
% Mfile 1
function [z]=Adams_Fourth_order(func,y0,x0,h,xf)
% z is the output
% func, y0, x0,h and xf are inputs
x=x0:h:xf;% Domain
y(1)=y0; % The initial value is used in the loop
for i=1:3
  k1=h*func(x(i),y(i));
   k2=h*func(x(i)+0.5*h,y(i)+0.5*k1);
   k3=h*func(x(i)+0.5*h,y(i)+0.5*k2);
   k4=h*func(x(i)+h,y(i)+k3);
   dely=(1/6)*(k1+2*k2+2*k3+k4);
   y(i+1)=y(i)+dely; %RK-4
end
for i=4:length(x)-1
  s=y(i)+(h/24)*(55*func(x(i),y(i))-...% Predictor
    59*func(x(i-1),y(i-1))+37*func(x(i-2),y(i-2))...
  -9*func(x(i-3),y(i-3)));
  ss=func(x(i+1),s); % Putting s in the function
  y(i+1)=y(i)+(h/24)*(9*ss+19*func(x(i),y(i))-...
  5*func(x(i-1),y(i-1))+func(x(i-2),y(i-2)));% Corrector
end
z=y;
```

```
% Mfile 2
function [z]=tys(x,y)
z=x*y;
end
```

```
>> % Command Window
>> Adams_Fourth_order(@tys,1,0,0.2,1)
```

6.1.5.3 Predictor-Corrector Method for the Systems of Ordinary Differential Equations

Example 6.12: Solve $y'' + 3y' - 2y = 5$ with $y(0) = 0$, $y'(0) = 1$ and h = 0.25, calculate $y(1) = ?$

Solution: Let

$$z = y',$$

so

$$z' = 5 - 3z + 2y,$$

with initial conditions

$$y(0) = 0, z(0) = 1.$$

We have $h = 0.25, x_0 = 0, y_0 = 0, z_0 = 1$.

Since $x_0 = 0, x_1 = x_0 + h = 0.2500, x_2 = x_0 + 2h = 0.5000,$

$x_3 = x_0 + 3h = 0.7500$ and $x_4 = x_0 + 4h = 1.0000$.

For i = 0

$$k_1 = hf(x_0, y_0, z_0) = hz_0 = (0.25)(1) = 0.2500,$$

$$l_1 = hg(x_0, y_0, z_0) = h(5 - 3z_0 + 2y_0)$$

$$= 0.25(5 - 3(1) + 2(0)) = 0.5000,$$

$$k_2 = hf(x_0 + 0.5h, y_0 + 0.5k_1, z_0 + 0.5l_1)$$

$$= h(z_0 + 0.5l_1) = 0.25(1 + 0.5(0.5000))$$

$$= 0.3125,$$

$$l_2 = hg(x_0 + 0.5h, y_0 + 0.5k_1, z_0 + 0.5l_1)$$

$$= h(5 - 3(z_0 + 0.5l_1) + 2(y_0 + 0.5k_1))$$

$$= 0.25(5 - 3(1 + 0.5(0.5000)) + 2(0 + 0.5(0.2500)))$$

$$= 0.3750,$$

$$k_3 = hf(x_0 + 0.5h, y_0 + 0.5k_2, z_0 + 0.5l_2)$$

$$= h(z_0 + 0.5l_2) = 0.25(1 + 0.5(0.3750))$$

$$= 0.2969,$$

$$l_3 = hg(x_0 + 0.5h, y_0 + 0.5k_2, z_0 + 0.5l_2)$$

$$= h(5 - 3(z_0 + 0.5l_2) + 2(y_0 + 0.5k_2))$$

$$= 0.25(5 - 3(1 + 0.5(0.3750)) + 2(0 + 0.5(0.3125)))$$

$$= 0.4375,$$

$$k_4 = hf(x_0 + h, y_0 + k_3, z_0 + l_3)$$

$$= h(z_0 + l_3) = 0.25(1 + 0.4375)$$

$$= 0.3594,$$

$$l_4 = hg(x_0 + h, y_0 + k_3, z_0 + l_3)$$

$$= h(5 - 3(z_0 + l_3) + 2(y_0 + k_3))$$

$$= 0.25(5 - 3(1 + 0.4375) + 2(0 + 0.2969))$$

$$= 0.3203,$$

$$y_1 = y_0 + \frac{1}{6}(k_1 + 2k_2 + 2k_3 + k_4)$$

$$= 0 + \frac{1}{6}(0.2500 + 2(0.3125) + 2(0.2969) + 0.3594)$$

$$= 0.3047,$$

$$z_1 = z_0 + \frac{1}{6}(l_1 + 2l_2 + 2l_3 + l_4)$$

$$= 1 + \frac{1}{6}(0.5000 + 2(0.3750) + 2(0.4375) + 0.3203)$$

$$= 1.4076.$$

By using the same procedure for i = 1 and 2, we get

$$y_2 = 0.6975, \ z_2 = 1.7262, \ y_3 = 1.1671 \text{ and } z_3 = 2.0305.$$

Now by using Adam's formula for i = 3, we will get:
Predictor:

$$y_{4,r} = y_3 + \left(\frac{h}{24}\right)\left[55y'_3 - 59y'_2 + 37y'_1 - 9y'_0\right],$$

$$y_{4,r} = 1.1671 + \left(\frac{0.25}{24}\right)\left[55(2.0305) - 59(1.7262) + 37(1.4076) - 9(1)\right],$$

$$y_{4,r} = 1.7182,$$

$$z_{4,r} = z_3 + \left(\frac{h}{24}\right)\left[55z'_3 - 59z'_2 + 37z'_1 - 9z'_0\right],$$

$$z_{4,r} = 2.0305 + \left(\frac{0.25}{24}\right)\left[55(1.2427) - 59(1.2165) + 37(1.3867) - 9(2)\right],$$

$$z_{4,r} = 2.3417.$$

By corrector formula:

$$y_{4,c} = y_3 + \left(\frac{h}{24}\right)\left[9y'_{4,r} + 19y'_3 - 5y'_2 + y'_1\right],$$

$$y_{4,c} = 1.1671 + \left(\frac{0.25}{24}\right)\left[9(2.3417) + 19(2.0305) - 5(1.7262) + (1.4076)\right],$$

$$y_{4,c} = 1.7132,$$

$$z_{4,c} = z_3 + \left(\frac{h}{24}\right)\left[9z'_{4,r} + 19z'_3 - 5z'_2 + z'_1\right],$$

$$z_{4,c} = 2.0305 + \left(\frac{0.25}{24}\right)\left[9(1.4112) + 19(1.2427) - 5(1.2165) + (1.3867)\right],$$

$$y_{4,c} = 2.3598.$$

Main steps for solving the predictor-corrector (the Adams-Bashforth formulas as a predictor and the Adams-Moulton formulas as a corrector) for the systems of ordinary differential equations in MATLAB: The output is z, the function 'func', the initial conditions y1 and x0, step size h and the final point of domain xf are inputs.

1. The domain is
   x = x0:h:xf.
2. The initial value y1 is represented by y(:,1) as a column vector.
3. The loop starts and the fourth order Runge-Kutta method for the system of ordinary differential equations is used to calculate the first three values of y.
4. Onward from 3, Adam's method is used and the formula for the predictor is written as
   s =y (:,i) + (h/24) * (55 * func(x(i),y(:,i)) – 59 * func(x(i – 1),y(:,i – 1)) + 37 * func(x(i – 2),y(:,i – 2)) …
   – 9 * func(x(i – 3),y(i – 3))).
5. This value is used in the function, i.e.,
   ss = func(x(i + 1),s).
6. The value ss is used in the corrector formula as
   y(:,i + 1) = y(:,i) + (h/24) * (9 * ss + 19 * func(x(i),y(:,i)) –- 5 * func(x(i –1),y(:,i – 1)) + … func((i – 2),y(:,i – 2))).
7. The function y′ is replaced with z.
8. The given function and Adam's method are run in the command window.

```
% Mfile 1
function [z]=Adams_Fourth_order_system(func,y1,x0,h,xf)
% z is the output
% func, y0, x0,h and xf are the inputs
x=x0:h:xf;% Domain
y(:,1)=y1; % Initial values
for i=1:3
  k1=h*func(x(i),y(:,i));
   k2=h*func(x(i)+0.5*h,y(:,i)+0.5*k1);
   k3=h*func(x(i)+0.5*h,y(:,i)+0.5*k2);
   k4=h*func(x(i)+h,y(:,i)+k3);
   dely=(1/6)*(k1+2*k2+2*k3+k4);
   y(:,i+1)=y(:,i)+dely; % RK four
end
for i=4:length(x)-1
  s=y(:,i)+(h/24)*(55*func(x(i),y(:,i))-...% Predictor
    59*func(x(i-1),y(:,i-1))...
  +37*func(x(i-2),y(:,i-2))...
  -9*func(x(i-3),y(:,i-3)))
  ss=func(x(i+1),s) % Putting s in the function
  y(:,i+1)=y(:,i)+(h/24)*... % Corrector
   (9*ss+19*func(x(i),y(:,i))-...
   5*func(x(i-1),y(:,i-1))+func(x(i-2),y(:,i-2)));
end
z=y';
```

```
% Mfile 2
function [z]=tys1(x,y)
z=[y(2); 5-3*y(2)+2*y(1)];
end
```

```
>> % Command Window
>> Adams_Fourth_order_system(@tys1,[0 1],0,0.25,1)
```

6.1.5.4   Milne's Method

Integrating Eq. (6.1) from $[x_{i-3}, x_{i+1}]$, we will get

$$y_{i+1} = y_{i-3} + \int_{x_{i-3}}^{x_{i+1}} f(x,y)\,dx. \tag{6.59}$$

The integral is the area under the curve form x_{i-3} to x_{i+1}. If $f(x,y)$ is approximated by the quadratic expression $ax^2 + bx + c$, then we can calculate the constants by using the values of the functions at $x_{i-3}, x_{i-2}, x_{i-1}, x_i$ and x_{i+1}. If the integral is calculated in such a way then we will get

$$y_{i+1,r} = y_{i-3} + \left(\frac{4h}{3}\right)\left[2f_i - f_{i-1} + 2f_{i-2}\right],$$

or

$$y_{i+1,r} = y_{i-3} + \left(\frac{4h}{3}\right)\left[2y'_i - y'_{i-1} + 2y'_{i-2}\right]. \tag{6.60}$$

Eq. (6.60) is known as the predictor for Milne's method. The corrector is achieved by using the Simpson's integration formula

$$y_{i+1,c} = y_{i-1} + \left(\frac{h}{3}\right)\left[f_{i+1} + 4f_i + f_{i-1}\right],$$

or

$$y_{i+1,c} = y_{i-1} + \left(\frac{h}{3}\right)\left[y'_{i+1} + 4y'_i + y'_{i-1}\right]. \tag{6.61}$$

Eq. (6.61) is known as the corrector for Milne's method. It is the fourth order method having a truncation error of order 4.

Note: The method of solving examples is similar to Adam's method, just a difference in formulas. So, codes are mentioned but examples are not solved.

```
% Mfile 1
function [z]=Milnes_Fourth_order(func,y0,x0,h,xf)
% z is the output
% func, y0,x0, h and xf are the inputs
x=x0:h:xf; % Domain
y(1)=y0; % The initial value is used in the loop
for i=1:3
  k1=h*func(x(i),y(i));
   k2=h*func(x(i)+0.5*h,y(i)+0.5*k1);
   k3=h*func(x(i)+0.5*h,y(i)+0.5*k2);
   k4=h*func(x(i)+h,y(i)+k3);
   dely=(1/6)*(k1+2*k2+2*k3+k4);
   y(i+1)=y(i)+dely; % RK four
end
for i=4:length(x)-1
  s=y(i-3)+ (4*h/3)*(2*func(x(i),y(i))-...% Predictor
    func(x(i-1),y(i-1))...
  +2*func(x(i-2),y(i-2)));
  ss=func(x(i+1),s); % Putting s in function
  y(i+1) =y(i-1) + (h/3)*...
```

```
      (ss+4*func(x(i),y(i))+...
      func(x(i-1),y(i-1))); % Corrector
end
z=y;
```

```
% Mfile 2
function [z]=tys(x,y)
z=x*y;
end
```

```
>> % Command Window
>> Milnes_Fourth_order(@tys,1,0,0.2,1)
```

```
% Mfile 1
function [z]=Milnes_Fourth_order_system(func,y1,x0,h,xf)
% z is the output
% func, y0, x0,h and xf are the inputs
x=x0:h:xf;% Domain
y(:,1)=y1; % Initial value
for i=1:3 % In order to make equal dimensions
  k1=h*func(x(i),y(:,i));
  k2=h*func(x(i)+0.5*h,y(:,i)+0.5*k1);
  k3=h*func(x(i)+0.5*h,y(:,i)+0.5*k2);
  k4=h*func(x(i)+h,y(:,i)+k3);
  dely=(1/6)*(k1+2*k2+2*k3+k4);
  y(:,i+1)=y(:,i)+dely; % RK four
end
for i=4:length(x)-1
  s=y(:,i-3)+(4*h/3)*(2*func(x(i),y(:,i))-...% Predictor
    func(x(i-1),y(:,i-1))...
  +2*func(x(i-2),y(:,i-2)));
  ss=func(x(i+1),s); % Putting s in the function
  y(:,i+1)=y(:,i-1)+(h/3)*...
    (ss+4*func(x(i),y(:,i))+...
    func(x(i-1),y(:,i-1))); % Corrector
end
z=y;
```

```
% Mfile 2
function [z]=tys1(x,y)
z=[y(2); 5-3*y(2)+2*y(1)];
end
```

```
>> % Command Window
>> Milnes_Fourth_order_system(@tys1,[0 1],0,0.25,1)
```

PROBLEM SET 6

1. Use the MATLAB codes of Euler's, Heun's and modified Euler's methods to calculate the solutions of the given problems with $h = 0.01$.

   (i) $y' = -x + y,\ (0 \le x \le 1),\ y(0) = 1.$

   (ii) $y' = y^4 - \left(\dfrac{4}{x}\right) y,\ (1 \le x \le 3), y(1) = 1.$

   (iii) $y' = \dfrac{(x^2 - xy + y^2)}{xy},\ (1 \le x \le 2), y(1) = 2.$

   (iv) $y' = \dfrac{2(1-y)}{x^2 \sin(x)},\ (1 \le x \le 2), y(1) = 2.$

2. Use the MATLAB codes of Euler's, Heun's and modified Euler's methods to calculate the solutions of the given systems with $h = 0.1$.

   (i) $y' = z, z' = -2z - y, y(0) = 2, z(0) = -1.$

   (ii) $y''' + 4y'' + 5y' = 0, y(0) = 1, y'(0) = 0, y''(0) = -1.$

   (iii) $y'' + 4y' + y = \sin(x), y(0) = 1, y'(0) = 0.$

3. Assume that the pendulum is fixed from a sliding collar (as shown in figure 6.7). The motion $y(t) = Y \sin(\kappa t)$ is executed on the collar when the system is at rest (at t = 0). The differential equation illustrating the motion of the pendulum is

$$\phi'' = -\frac{g}{l}\sin(\phi) + \frac{\kappa^2}{l}Y\cos(\phi)\sin(\kappa t).$$

Use $l = 1\text{m}$, g = 9.8 $m/s^2, Y = 0.21$ and $\kappa = 2.1 rad/s$ to calculate the solution of the problem by classical fourth order Runge-Kutta code. Where $0 \le t \le 10$.

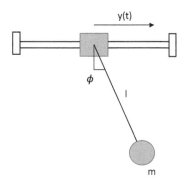

FIGURE 6.7   Fixed pendulum.

4. Use the classical fourth order Runge-Kutta code for the systems of equations to approximate the solutions of the following initial value problems ($0 \leq x \leq 3$).

(i) $y''' + yy'' + y'^2 = 0, y(0) = 0, y'(0) = 1, y''(0) = 0.1.$

(ii) $y'' + yy'^2 = 0, y(0) = 0, y'(0) = 0.1.$

(iii) $y''' + 0.5yy'' = 0, y(0) = 0, y'(0) = 0, y''(0) = 1.$

5. Use Runge-Kutta-Fehlberg code with h = 0.1 and tolerance $\varepsilon = 0.001$ to calculate the solution of the following problems:

(i) $y' = y - x^2 + 1, 0 \leq x \leq 3, y(0) = 1.$

(ii) $y' = (x + 2x^3)y^3 - xy, 0 \leq x \leq 3, y(0) = 0.3.$

(iii) $y' = -(y+1)(y+3), 0 \leq x \leq 3, y(0) = 4.$

6. Use Runge-Kutta-Fehlberg code for the system of equations with h = 0.1 and tolerance $\varepsilon = 0.001$ to calculate the solution of the following problems ($0 \leq x \leq 3$):

(i) $y''' - y'\ln(y'') - sin(y) = 0, y(0) = 0, y'(0) = 1, y''(0) = 1.$

(ii) $y'' = -(x^2 - 1)y' - y, y(0) = 0.4, y'(0) = 1.$

(iii) $y'' + 4y' + y = sin(y), y(0) = 1, y'(0) = 0.1.$

7. Assume a force G(t) is applied to a mass-spring system which is at rest (see Figure 6.8)

$$G(t) = \begin{bmatrix} 5tN & \text{when } t < 1s, \\ 10tN & \text{when } t > 1s \end{bmatrix}.$$

The differential equation for the problem is

$$y'' = \frac{G(t)}{m} - \frac{f}{m}y.$$

Calculate the maximum displacement of the mass by using the predictor-corrector method where m = 1.2kg and $f = 21N / m$.

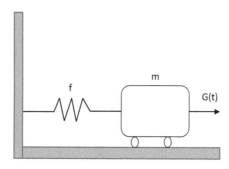

FIGURE 6.8   Mass-spring system.

8. Use predictor-corrector method code for the system of equations with h = 0.1 to calculate the solutions of the following problems $(0 \leq x \leq 3)$:

(i) $y''' + 0.2\left(yy'' + 2y'^2\right) = 0, y(0) = 0, y'(0) = 1, y''(0) = 1.2.$

(ii) $y'' + e^y\left(1 + y'\right) = 0, y(0) = ln(0.1), y'(0) = 1.$

7 Boundary Value Problems (BVPs)

As in the previous chapter, the equations were solved by conditions imposed at one point (initial value). In this chapter, the approximate solutions are calculated by conditions imposed at more than one point (boundary value).

Consider a heated rod (as shown in Figure 7.1) of length L which is assumed to be insulated and in steady state results in

$$\frac{d^2T}{dx^2} + k\left(T_s - T\right) = 0, \tag{7.1}$$

where k is the heat transfer coefficient and T_s is the temperature of the surrounding air.

The temperature at the ends of the rod is held fixed, i.e.,

$$T(0) = T_0, T(L) = T_1. \tag{7.2}$$

Eq. (7.1) along with conditions (7.2) is an example of the boundary value problem.

In this book, two numerical approaches are used to calculate the solution of the linear and non-linear boundary value problems, namely the shooting method and the finite difference method.

7.1 LINEAR SHOOTING METHOD

Consider a second order linear boundary value problem

$$y'' = A(x)y' + B(x)y + C(x),\ a \le x \le b, y(a) = \alpha_1, y(b) = \alpha_2. \tag{7.3}$$

Where $A(x), B(x)$ and $C(x)$ are continuous and $B(x) > 0$ on [a,b]. The linear shooting method depends on converting the linear boundary value problem into two initial value problems, i.e.,

$$y'' = A(x)y' + B(x)y + C(x),\ a \le x \le b, y(a) = \alpha_1, y'(a) = 0, \tag{7.4}$$

and

$$y'' = A(x)y' + B(x)y,\ a \le x \le b, y(a) = 0, y'(a) = 1. \tag{7.5}$$

Define

$$y(x) = y_1(x) + \frac{\alpha_2 - y_1(b)}{y_2(b)} y_2(x). \tag{7.6}$$

Eq. (7.6) is the solution of the linear boundary value problem (7.3). The solutions $y_1(x)$ and $y_2(x)$ are calculated by using any initial value method, which were discussed in the previous chapter, and

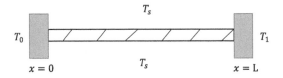

FIGURE 7.1 A graphical behavior of a thin rod.

DOI: 10.1201/9781003385288-7

once these approximations are calculated the results are substituted in Eq. (7.6) to get the solution of the boundary value problem. The graphical behavior is mentioned in Figure 7.2.

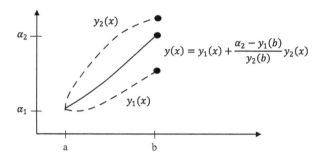

FIGURE 7.2   A graphical behavior of linear shooting method.

Example 7.1: Use the shooting method to solve the numerical solution of the problem $y'' = 2y - (x+1)y' + (1-x^2)e^{-x}$ with $y(0) = 0$ and $y(1) = 0$.

Solution: Convert the boundary value problem into two initial value problems

$$y'' = 2y - (x+1)y' + (1-x^2)e^{-x}, \quad y(0) = 0, y'(0) = 0,$$

$$y'' = 2y - (x+1)y', \quad y(0) = 0, y'(0) = 1.$$

As discussed in the previous chapter, the given initial value problem is converted into four initial value systems. By using the fourth order Runge-Kutta method solutions are calculated. For the fourth order Runge-Kutta method the values of k_1, k_2, k_3 and k_4 (for i = 0, 1, 2, 3 and 4) are (see Table 7.1):

TABLE 7.1
The Numerical Values of k's

i	k_1	k_2	k_3	k_4
0	0	0.0200	0.0157	0.0332
	0.2	0.1572	0.1659	0.1237
	0.2	0.1800	0.1842	0.1667
	–0.2	–0.1580	–0.1666	–0.1263
1	0.0323 0.1254 0.1675	0.0449	0.0413	0.0517
	–0.1280	0.0899	0.0971	0.0637
		0.1547	0.1580	0.1472
		–0.0946	–0.1015	–0.0698
2	0.0511	0.0576	0.0550	0.0599
	0.0652	0.0389	0.0441	0.0204
	0.1478	0.1407	0.1432	0.1376
	–0.0713	–0.0458	–0.0510	–0.0272
3	0.0595	0.0616	0.0599	0.0610
	0.0216	0.0043	0.0076	–0.0070
	0.1381	0.1352	0.1371	0.1354
	–0.0284	–0.0097	–0.0135	0.0037
4	0.0608	0.0601	0.0592	0.0579
	–0.0063	–0.0159	–0.0142	–0.0214
	0.1357	0.1360	0.1373	0.1384
	0.0028	0.0159	0.0134	0.0252

By using the fourth order Runge-Kutta method the solutions are

$$y_1(0) = 0, y_1(0.2) = 0.0174,$$

$$y_1(0.4) = 0.0602,$$

$$y_1(0.6) = 0.0602, y_1(0.8) = 0.1768,$$

$$y_1(1) = y_1(b) = 0.2363.$$

$$y_2(0) = 0, y_2(0.2) = 0.1825,$$

$$y_2(0.4) = 0.3392,$$

$$y_2(0.6) = 0.4814, y_2(0.8) = 0.6178,$$

$$y_2(1) = y_2(b) = 0.7545.$$

Now by using the formula Eq. (7.6) the values are calculated for 0.2, 0.6, 0.8 and 1.
For $x = 0.2$, we have

$$y(0.2) = y_1(0.2) + \frac{0 - y_1(b)}{y_2(b)} y_2(0.2)$$

$$= 0.0174 + \frac{0 - 0.2363}{0.7545} 0.1825$$

$$= -0.0398.$$

By using the same procedure, the remaining values are

$$y(0.4) = -0.0461, y(0.6)$$

$$= -0.0346, y(0.8)$$

$$= -0.0167 \text{ and } y(1) = 0.$$

Main steps for solving the shooting method for a linear ordinary differential equation in MATLAB: The output is denoted by z, the inputs are the function 'func', the initial condition x0, the step size h and final point of domain xf.

1. For the linear shooting method, the boundary value problem is converted into a set of initial value problems. The systems are solved by using any initial value method. For the second order boundary value problem, four initial value systems are generated. So, four initial values are required, i.e.,
   y1 = [0 0 0 1].
2. Define the domain
   x = x0:h:xf.
3. The initial guess in column form will be
   y(:,1) = y1.
4. Now the fourth order Runge-Kutta method is used to find the solution. The loop starts from i = 1:length(x) – 1, with formulas
   k1 = h * func(x(i),y(:,i)), k2 = h * func(x(i) + 0.5 * h,y(:,i) + 0.5 * k1), k3 = h * func(x(i) + 0.5 * h,y(:,i) + 0.5 * k2), k4 = h * func(x(i) + h,y(:,i) + k3), dely = (1/6) * (k1 + 2 * k2 + 2 * k3 + k4), y(:,i + 1) = y(:,i) + dely.

5. As seen in the formula Eq. (7.6), the solutions $y_1(x)$ and $y_2(x)$ along with their end values are used to calculate the final solution, i.e.,
   a = y(1,end), b = y(3,end), z = y(1,:) – (a/b) * y(3,:).
6. Use the second Mfile to write the given problem and the command window is used to run both files.

```
% Mfile 1
function [z]=Linear_shooting_single(func,x0,h,xf)
% z is the output
% func, y1, x0,h and xf are the inputs
y1=[0 0 0 1];
x=x0:h:xf; % Domain
y(:,1)=y1; % One column
for i=1:length(x)-1
k1=h*func(x(i),y(:,i));
k2=h*func(x(i)+0.5*h,y(:,i)+0.5*k1);
k3=h*func(x(i)+0.5*h,y(:,i)+0.5*k2);
k4=h*func(x(i)+h,y(:,i)+k3);
dely=(1/6)*(k1+2*k2+2*k3+k4);
y(:,i+1)=y(:,i)+dely; % Rk4 formula
end
a=y(1,end);
b=y(3,end);
z=y(1,:)-(a/b)*y(3,:);    % Linear shooting formula
```

```
% Mfile 2
function [z]=Linear_shooting_example(x,y)
z=[y(2); 2*y(1)-(x+1)*y(2)+(1-x^2)*exp(-x); y(4);2*y(3)-(x+1)*y(4)];
end
```

```
>> % Command Window
>> Linear_shooting_single(@Linear_shooting_example,0,0.2,1)
```

7.2 SHOOTING METHOD FOR THE LINEAR ORDINARY DIFFERENTIAL SYSTEMS

Consider a system of linear boundary value problems

$$y'' = A(x)y' + B(x)y + C(x),$$
$$a \le x \le b, y(a) = \alpha_1, y(b) = \alpha_2, \tag{7.7}$$

$$g'' = D(x)g' + E(x)g + F(x),$$
$$a \le x \le b, g(a) = \beta_1, g(b) = \beta_2. \tag{7.8}$$

Converting the linear boundary value problem into an initial value problem, we get

$$y'' = A(x)y' + B(x)y + C(x),$$
$$a \le x \le b, y(a) = \alpha_1, y'(a) = 0,$$

(7.9)

$$y'' = A(x)y' + B(x)y,$$
$$a \le x \le b, y(a) = 0,$$
$$y'(a) = 1,$$

(7.10)

$$g'' = D(x)g' + E(x)g + F(x),$$
$$a \le x \le b, g(a) = \beta_1,$$
$$g'(a) = 0,$$

(7.11)

$$g'' = D(x)g' + E(x)g, a \le x \le b,$$
$$g(a) = 0, g'(a) = 1.$$

(7.12)

Define

$$y(x) = y_1(x) + \frac{\alpha_2 - y_1(b)}{y_2(b)} y_2(x),$$

(7.13)

$$g(x) = g_1(x) + \frac{\beta_2 - g_1(b)}{g_2(b)} g_2(x).$$

(7.14)

Eqs. (7.13) and (7.14) are the solutions of the linear boundary value problems Eqs. (7.7) and (7.8).

Example 7.2: Use the shooting method to solve the system of equations $y'' = y + y' + 1, g'' = g + g' + 3,$ with boundary conditions $y(0) = 0, g(0) = 0, y(1) = 0$ and $g(1) = 0$.

Solution: Convert the boundary value problem into an initial value problem

$$y'' = y + y' + 1, y(0) = 0, y'(0) = 0,$$

$$y'' = y + y', y(0) = 0, y'(0) = 1,$$

$$g'' = g + g' + 3, g(0) = 0, g'(0) = 0,$$

$$g'' = g + g', g(0) = 0, g'(0) = 1.$$

The initial value problem is converted into eight first order systems. By using the fourth order Runge-Kutta method the values of k_1, k_2, k_3 and k_4 (for i = 0, 1, 2, 3 and 4) are (see Table 7.2):

TABLE 7.2
The Numerical Values of k's

i	k_1	k_2	k_3	k_4
0	0	0.0200	0.0220	0.0448
	0.2000	0.2200	0.2240	0.2492
	0.2000	0.2200	0.2240	0.2492
	0.2000	0.2400	0.2460	0.2940
	0	0.0600	0.0660	0.1344
	0.6000	0.6600	0.6720	0.7476
	0.2000	0.2200	0.2240	0.2492
	0.2000	0.2400	0.2460	0.2940
1	0.0446	0.0695	0.0724	0.1013
	0.2489	0.2782	0.2836	0.3201
	0.2489	0.2782	0.2836	0.3201
	0.2934	0.3477	0.3560	0.4214
	0.1337	0.2084	0.2172	0.3039
	0.7466	0.8346	0.8509	0.9602
	0.2489	0.2782	0.2836	0.3201
	0.2934	0.3477	0.3560	0.4214
2	0.1010	0.1330	0.1372	0.1748
	0.3196	0.3617	0.3691	0.4209
	0.3196	0.3617	0.3691	0.4209
	0.4206	0.4946	0.5062	0.5957
	0.3030	0.3989	0.4115	0.5244
	0.9588	1.0850	1.1072	1.2626
	0.3196	0.3617	0.3691	0.4209
	0.4206	0.4946	0.5062	0.5957
3	0.1744	0.2164	0.2224	0.2724
	0.4202	0.4797	0.4898	0.5626
	0.4202	0.4797	0.4898	0.5626
	0.5946	0.6961	0.7122	0.8350
	0.5232	0.6492	0.6671	0.8171
	1.2606	1.4390	1.4694	1.6879
	0.4202	0.4797	0.4898	0.5626
	0.5946	0.6961	0.7122	0.8350
4	0.2718	0.3280	0.3363	0.4036
	0.5617	0.6451	0.6590	0.7608
	0.5617	0.6451	0.6590	0.7608
	0.8335	0.9731	0.9953	1.1644
	0.8154	0.9839	1.0089	1.2108
	1.6852	1.9353	1.9771	2.2824
	0.5617	0.6451	0.6590	0.7608
	0.8335	0.9731	0.9953	1.1644

The solutions are

$$y_1(0) = 0, y_1(0.2) = 0.0215, y_1(0.4) = 0.0931$$

$$y_1(0.6) = 0.2291, y_1(0.8) = 0.4498,$$

$$y_1(1) = y_1(b) = 0.7838.$$

$$y_2(0) = 0, y_2(0.2) = 0.2229,$$

$$y_2(0.4) = 0.5050, y_2(0.6) = 0.8720,$$

$$y_2(0.8) = 1.3589,$$

$$y_2(1) = y_2(b) = 2.0141.$$

$$g_1(0) = 0, g_1(0.2) = 0.0644, g_1(0.4) = 0.2792,$$

$$g_1(0.6) = 0.6872, g_1(0.8) = 1.3494,$$

$$g_1(1) = g_1(b) = 2.3513.$$

$$g_2(0) = 0, g_2(0.2) = 0.2229, g_2(0.4) = 0.5050,$$

$$g_2(0.6) = 0.8720, g_2(0.8) = 1.3589,$$

$$g_2(1) = g_2(b) = 2.0141.$$

By using the formulas Eqs. (7.13) and (7.14), the values are calculated for 0.2, 0.6 and 0.8. For $x = 0.2$, we have

$$y(0.2) = y_1(0.2) + \frac{0 - y_1(b)}{y_2(b)} y_2(0.2)$$

$$= 0.0215 + \frac{0 - 0.7838}{2.0141} 0.2229$$

$$= -0.0652,$$

$$g(0.2) = g_1(0.2) + \frac{0 - g_1(b)}{g_2(b)} g_2(0.2)$$

$$= 0.0644 + \frac{0 - 2.3513}{2.0141} 0.2229$$

$$= -0.1958.$$

By using the same procedure, the remaining values are

$$y(0.4) = -0.1034, y(0.6)$$

$$= -0.1103, y(0.8)$$

$$= -0.0790 \text{ and } y(1) = 0,$$

$$g(0.4) = -0.3103, g(0.6)$$

$$= -0.3308, g(0.8)$$

$$= -0.2371 \text{ and } g(1) = 0.$$

Main steps for solving the shooting method for the system of linear ordinary differential equations in MATLAB: The outputs are denoted by z and s, the inputs are the function 'func', the initial condition x0, the step size h and the final point of domain xf.

1. By converting the boundary value problem into a set of initial value problems eight initial conditions are needed, which will be
   y1=[0 0 0 1 0 0 0 1].

2. The domain is defined by
   x = x0:h:xf.
3. The initial guess in column form is
   y(:,1) = y1.
4. Now, the fourth order Runge-Kutta method is used to find the solution.
5. By using the formulas Eqs. (7.13) and (7.14), the solutions $y_1(x)$, $y_2(x)$, $g_1(x)$ and $g_2(x)$, along with their end values, are used to calculate the final form, i.e.,

```
a = y(1,end),
b = y(3,end),
c = y(5,end),
d = y(7,end),
z = y(1,:) - (a/b) * y(3,:),
s = y(5,:) - (c/d) * y(7,:).
```

6. The second Mfile is used to write the given problem, i.e.,
   z = [y(2); y(1) + y(2) + 1; y(4); y(3) + y(4); y(6); y(5) + y(6) + 3; y(8); y(7) + y(8)].
7. The command window is used to run both files.

```
% Mfile 1
function [z,s]=Linear_shooting_systems(func,x0,h,xf)
% z is the output
% func, y1, x0,h and xf are the inputs
y1=[0 0 0 1 0 0 0 1];
x=x0:h:xf; % Domain
y(:,1)=y1; % One column
for i=1:length(x)-1
k1=h*func(x(i),y(:,i));
   k2=h*func(x(i)+0.5*h,y(:,i)+0.5*k1);
   k3=h*func(x(i)+0.5*h,y(:,i)+0.5*k2);
   k4=h*func(x(i)+h,y(:,i)+k3);
   dely=(1/6)*(k1+2*k2+2*k3+k4);
   y(:,i+1)=y(:,i)+dely % Rk4 formula
end
a=y(1,end);
b=y(3,end);
z=y(1,:)-(a/b)*y(3,:);  % Linear shooting formula
c=y(5,end);
d=y(7,end);
s=y(5,:)-(c/d)*y(7,:);  % Linear shooting formula
```

```
% Mfile 2
function [z]=Linear_shooting_example_systmes(x,y)
z=[y(2); y(1)+y(2)+1; y(4);y(3)+y(4); y(6); y(5)+y(6)+3; y(8);
y(7)+y(8)];
end
```

```
>> % Command Window
>> [z,s]=Linear_shooting_systems(@Linear_shooting_example_systmes,0,
0.2,1)
```

7.3 SHOOTING METHOD FOR NONLINEAR ORDINARY DIFFERENTIAL EQUATIONS

Consider a second order nonlinear boundary value problem

$$y'' = f(x, y, y'), y(a) = A, y(b) = B. \tag{7.15}$$

The equation is converted into an initial value problem. For this $y'(a)$ is assumed with some random value.

$$y'' = f(x, y, y'), y(a) = A, y'(a) = u. \tag{7.16}$$

The correct value of u gives the desired result. If the achieved solution's final value and the boundary condition are approximately equal, then we are done. Otherwise, we must choose some different initial guess and try again and again until we get the desired result (as shown in Figure 7.2). The problem is to find u, such that

$$y(u, b) - B = 0. \tag{7.17}$$

Different root finding methods are used to calculate the correct value of the initial guess. In this book, secant and the Newton Raphson methods are used to calculate the initial guess. If we use the secant method for nonlinear equations, then we will use the initial value method twice for initial guesses and the remaining procedure is controlled by a loop in which we use the secant method again with any initial value method. For the Newton Raphson method, the initial value problem is used once before the loop, and the remaining procedure is controlled by a loop in which we use the Newton Raphson method again with any initial value method (see Figure 7.3).

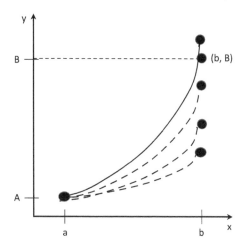

FIGURE 7.3 A graphical behavior of nonlinear shooting method.

7.3.1 SECANT METHOD COMBINED WITH EULER'S METHOD: FOR SECOND ORDER NONLINEAR PROBLEMS

Consider a second order nonlinear boundary value problem

$$y'' = f(x, y, y'), y(a) = A, y(b) = B. \tag{7.18}$$

Convert the boundary value problem into an initial value problem

$$y'' = f(x, y, y'), y(a) = A, y'(a) = u_1. \tag{7.19}$$

As $y'(a) = u_1$ is unknown, we choose any arbitrary value of u_1 and solve the initial value problem by using Euler's method (as discussed in Chapter 6). Let $y(u_1, b)$ be the computed final value of the solution at $x = b$. Our focus is to calculate u_1 such that

$$y(u_1, b) - B = 0. \tag{7.20}$$

Again, any arbitrary value of u_2 is chosen and the same procedure is repeated. Let $y(u_2, b)$ be the computed final value of the solution at $x = b$ by using Euler's method. Now the next initial guess is calculated by using these two values in the secant method, which gives

$$u_3 = u_2 - (u_2 - u_1) \frac{y(u_2, b) - B}{y(u_2, b) - y(u_1, b)}. \tag{7.21}$$

The value u_3 is used again in Euler's method. And the procedure is repeated until we get the desired boundary condition. The general form of the secant method is

$$u_{i+1} = u_i - (u_i - u_{i-1}) \frac{y(u_i, b) - B}{y(u_i, b) - y(u_{i-1}, b)}. \tag{7.22}$$

Note that in this method a good choice of u_1 and u_2 is essential.

Example 7.3: Use the shooting method (Euler's and secant method) to calculate the approximate solution of the nonlinear boundary value problem

$$y'' = -\frac{(y')^2}{y},$$

$$y(1) = \sqrt{2}, y(3) = 2.$$

Solution: First we need to convert the boundary value problem into an initial value problem.
   Let

$$z = y',$$

so

$$z' = -\frac{(z)^2}{y}.$$

With initial conditions

$$y(1) = \sqrt{2}, z(1) = u_1 = 1.3.$$

We have $h = 0.5, x_0 = 1, y_0 = \sqrt{2}$, $z_0 = 1.3$.
   Since $x_0 = 1, x_1 = x_0 + h = 1.5, x_2 = x_0 + 2h = 2,$
      $x_3 = x_0 + 3h = 2.5, x_4 = x_0 + 4h = 3.$
   Here

$$f = z,$$

$$g = -\frac{(z)^2}{y}.$$

Using Euler's formula for i = 0, 1, 2 and 3

$$y_1 = 2.0642, y_2 = 2.4155, y_3 = 2.7069, y_4 = 2.9632.$$

$$z_1 = 0.7025, z_2 = 0.5830, z_3 = 0.5126, z_4 = 0.4641.$$

As there is a difference between the boundary condition which is $y(3) = 2$ and the calculated value $y_4 = 2.9632$. The difference is

$$a(1) = y(1.3,3) - 2 = 2.9632 - 2 = 0.9632.$$

Again, the same procedure is repeated by choosing a different initial guess. Assume

$$z_0 = u_2 = 0.3.$$

Using Euler's formula for i = 0, 1, 2 and 3

$$y_1 = 1.5642, y_2 = 1.6983, y_3 = 1.8209, y_4 = 1.9346.$$

$$z_1 = 0.2682, z_2 = 0.2452, z_3 = 0.2275, z_4 = 0.2133.$$

Again, there is a difference between the boundary condition which is $y(3) = 2$ and the calculated value $y_4 = 1.9346$. The difference is

$$a(2) = y(0.3,3) - 2 = 1.9346 - 2 = -0.0654.$$

Now the initial guesses (u_1, u_2) and the final values $(a1, a2)$ are utilized in the secant method as:

$$u_{i+1} = u_i - \frac{(u_i - u_{i-1})a(i)}{a(i) - a(i-1)}.$$

For i = 2, we get

$$u_3 = u_2 - \frac{(u_2 - u_1)a(2)}{a(2) - a(1)}$$

$$= 0.3 - \frac{(0.3 - 1.3)(-0.0654)}{-0.0654 - 0.9632} = 0.3635.$$

This value will be our next initial value and the same procedure is repeated by using Euler's method with $z_0 = u_3 = 0.3635$.

Using Euler's formula for i = 0, 1, 2 and 3

$$y_1 = 1.5960, y_2 = 1.7544, y_3 = 1.8971,$$

$$y_4 = 2.0282,$$

$$z_1 = 0.3168, z_2 = 0.285,$$

$$z_3 = 0.2622, z_4 = 0.2440.$$

The difference between the boundary condition $y(3) = 2$ and the calculated value $y_4 = 2.0282$ is

$$a(3) = 2.0282 - 2 = 0.0282.$$

Again, the difference is large, use the secant method again for i = 3, we get $u_4 = 0.3444$.

This value will be our next initial value and the same procedure is repeated by using Euler's method with

$$z_0 = u_4 = 0.3444.$$

Using Euler's formula for i = 0, 1, 2 and 3, we get

$$y_1 = 1.5864, y_2 = 1.7377, y_3 = 1.8745, y_4 = 2.0005.$$

$$z_1 = 0.3025, z_2 = 0.2736, z_3 = 0.2521, z_4 = 0.2351.$$

As we can see, the difference between the boundary condition $y(3) = 2$ and the calculated value $y_4 = 2.0005$ is small. So, we can stop here or can refine the guess by using the secant method again to get a more accurate result.

The procedure is stopped here, and the above values of y and z are the required solution of the boundary value problem.

Main steps for solving the secant method combined with the Euler's method (shooting method for second order nonlinear ordinary differential equation) in MATLAB: The output will be z, the inputs are the function 'func', initial condition x0, step size h and final point of domain xf.

1. First, we use the previous procedure of Euler's method for the system of equations, we have one known initial value and we choose the guess randomly, i.e.,
   u1 = 1.3,
   y1 = [sqrt(2) u1].
   This will give us a row with two values:
   y1 = 1.4142 1.3000.
2. Define the domain
   x = x0:h:xf.
3. As y1 is a row vector, we must convert it into a column vector as the system is defined in column form, i.e.,
   y(:,1) = y1.
4. Now Euler's method is used along with the given initial conditions. The loop starts from i = 1:length(x) – 1, with formula
   y(:,i + 1) = y(:,i) + h * func(x(i),y(:,i)).
5. In column form we can write the results as
   t = y'.
6. The important step is to take the last value from the first column and subtract it from the desired boundary condition, this will be our first function value which will be used in the secant method.
   a1 =t (end,1) – 2 (a1 = last value of the first column – the boundary condition).
7. Again, a different initial guess (u2 = 0.3) is assumed, and the same procedure is repeated for the second value of the secant method.
   a2 = t(end,1) – 2.
8. A loop is generated for the refinement in initial guesses by the secant method.
9. The calculated values of initial guesses (u1,u2) and functions (a1,a2) are defined by
   s = a1,
   r=a2,
   q=u1,
   u=u2,
   with the secant formula
   u3 = u – ((u – q)/(r – s)) * (r).

This is the most important step, as we can see this procedure is adapted because the values of initial guesses (u1,u2) and functions (a1,a2) are refined in each iteration.

The loop is not stopped here; we use another loop for Euler's method and the value of u3 is used in Euler's method, which gives

v = t(end,1) – 2.

10. The values of the secant formula are refined by

a1 = r,
a2 = v,
u1 = u,
u2 = u3.

11. This will refine the values in the secant method and a new initial guess is achieved and the guess is refined until it satisfies the difference between the calculated value and the boundary condition.

12. The second Mfile is used to write the given second order problem

z = [y(2); –(y(2) ^ 2)/y(1)].

13. Using the command window both files are run, and the required result is calculated.

```
% Mfile 1
function [z]=Euler_secant(func,x0,h,xf)
% z is the output
% func, x0,h and xf are the inputs
u1=1.3;
y1=[sqrt(2) u1];
x=x0:h:xf; % Domain
y(:,1)=y1; % One column
for i=1:length(x)-1
  y(:,i+1)=y(:,i)+h*func(x(i),y(:,i)); % The Euler's formula
end
t=y'; % Column vector
a1=t(end,1)-2; % Last value of the first column - the boundary
% condition
for numiter=1:6
  s=a1;
  r=a2;
  q=u1;
  u=u2;
  u3=u -((u-q)/(r-s))*(r); % Secant formula
for i=1:length(x)-1
y1=[(2)^(1/2) u3];
 x=x0:h:xf; % Domain
y(:,1)=y1; % One column
y(:,i+1)=y(:,i)+h*func(x(i),y(:,i)); % Euler's formula
end
t=y';
v=t(end,1)-2; % Last value of the first column - the boundary
% condition
a1=r; % Refined values
a2=v; % Refined values
u1=u; % Refined values
u2=u3; % Refined values
end
z=t;
plot(x,z(:,1))
end
```

```
% Mfile 2
function [z]=example(x,y)
z=[y(2); -(y(2)^2)/y(1)];
end
```

```
>> % Command Window
>> Euler_secant (@example,1,0.5,3)
```

7.3.2 SECANT METHOD COMBINED WITH EULER'S METHOD: FOR COUPLED NONLINEAR ORDINARY DIFFERENTIAL EQUATIONS

Let us consider a boundary value system

$$y''' = f(x, y, y', y'', w, w'), y(a) = A, y'(a) = B, y(b) = C, \tag{7.23}$$

$$w'' = g(x, y, y', w, w'), w(a) = E, w(b) = F. \tag{7.24}$$

The system is converted into an initial value problem

$$y''' = f(x, y, y', y'', w, w'), y(a) = A, y'(a) = B, y''(a) = u_1, \tag{7.25}$$

$$w'' = g(x, y, y', w, w'), w(a) = E, w'(a) = v_1. \tag{7.26}$$

As $y''(a) = u_1$ and $w'(a) = v_1$ are unknown, we choose any arbitrary values of u_1 and v_1 and solve the initial value problem by using Euler's method. We must calculate u_1 and v_1 such that

$$y'(u_1, b) - C = 0, \tag{7.27}$$

$$w(v_1, b) - F = 0. \tag{7.28}$$

Again, the procedure is repeated for arbitrary values of u_2 and v_2. Let $y'(u_2, b)$ and $w(v_2, b)$ be the computed final values of the solution at $x = b$ by using Euler's method. Next, initial guesses are calculated by using these values in the secant method

$$u_3 = u_2 - (u_2 - u_1) \frac{y'(u_2, b) - C}{y'(u_2, b) - y'(u_1, b)}, \tag{7.29}$$

$$v_3 = v_2 - (v_2 - v_1) \frac{w(v_2, b) - F}{w(v_2, b) - w(v_1, b)}. \tag{7.30}$$

These values are used again in Euler's method. And the procedure is repeated until we get the desired boundary conditions.

Example 7.4: Use the shooting method (Euler's and secant method) to calculate the approximate solution of the nonlinear boundary value problem

$$y''' + yy'' - y'^2 = 0,$$

$$w'' - y'w = 0,$$

$$y(0) = 0, y'(0) = 1, w(0) = 1, y'(1) = 0, w(1) = 0.$$

Solution: First we convert the boundary value problem into the initial value problem.
Let

$$y' = z,$$

$$z' = y'' = t,$$

$$t' = -yt + z^2,$$

$$w' = l,$$

$$l' = zw.$$

With initial conditions

$$y(0) = 0, z(0) = 1, t(0) = u_1 = 1.3, w(0) = 0, l(0) = v_1 = 1.2.$$

We have $h = 0.25, x_0 = 0, y_0 = 0, z_0 = 1, t_0 = 1.3, l_0 = 1.2$.
Since $x_0 = 0, x_1 = x_0 + h = 0.25, x_2 = x_0 + 2h = 0.5,$

$x_3 = x_0 + 3h = 0.75, x_4 = x_0 + 4h = 1$.
Now by applying Euler's method for i = 0, 1, 2, 3

$$y_1 = 0.2500, y_2 = 0.5813, y_3 = 1.0094, y_4 = 1.5558.$$

$$z_1 = 1.3250, z_2 = 1.7125, z_3 = 2.1855, z_4 = 2.7731.$$

$$t_1 = 1.5500, t_2 = 1.8920, t_3 = 2.3503, t_4 = 2.9513.$$

$$w_1 = 1.3000, w_2 = 1.6625, w_3 = 2.1327, w_4 = 2.7808.$$

$$l_1 = 1.4500, l_2 = 1.8806, l_3 = 2.5924, l_4 = 3.7576.$$

As there is a difference between the boundary conditions which are $y'(1) = z(1) = 0$ and $w(1) = 0$ and the calculated values $z_4 = 2.7731$ and $w_4 = 2.7808$. The difference is

$$a(1) = y'(1.3,1) - 0 = 2.7731 - 0 = 2.7731,$$

$$b(1) = w(1.3,1) - 0 = 2.7808 - 0 = 2.7808.$$

Again, the same procedure is repeated by choosing different initial guesses. Assume

$$t(0) = u_2 = 0.3, l(0) = v_2 = 0.2.$$

Now by applying Euler's method again.
For i = 0, 1, 2 and 3

$$y_1 = 0.2500, y_2 = 0.5188, y_3 = 0.8219, y_4 = 1.1753.$$

$$z_1 = 1.0750, z_2 = 1.2125, z_3 = 1.4136, z_4 = 1.6806.$$

$$t_1 = 0.5500, t_2 = 0.8045, t_3 = 1.0677, t_4 = 1.3479.$$

$$w_1 = 1.0500, w_2 = 1.1625, w_3 = 1.3455, w_4 = 1.6167.$$

$$l_1 = 0.4500, l_2 = 0.7322, l_3 = 1.0846, l_4 = 1.5601.$$

As there is a difference between the boundary conditions which are $y'(1) = z(1) = 0$ and $w(1) = 0$ and the calculated values $z_4 = 1.6806, w_4 = 1.6167$. The difference is

$$a(2) = y'(0.3,1) - 0 = 1.6806 - 0 = 1.6806,$$

$$b(2) = w(0.3,1) - 0 = 1.6167 - 0 = 1.6167.$$

The initial guesses (u_1, u_2, v_1, v_2) and the final values $(a(1), a(2), b(1), b(2))$ are utilized in the secant method as:

$$u_{i+1} = u_i - \frac{(u_i - u_{i-1})a(i)}{a(i) - a(i-1)},$$

$$v_{i+1} = v_i - \frac{(v_i - v_{i-1})b(i)}{b(i) - b(i-1)}.$$

For i = 2, we get

$$u_3 = u_2 - \frac{(u_2 - u_1)a(2)}{a(2) - a(1)}$$

$$= 0.3 - \frac{(0.3 - 1.3))1.6806}{1.6806 - 2.7731}$$

$$= -1.2383,$$

$$v_3 = v_2 - \frac{(v_2 - v_1)b(2)}{b(2) - b(1)} = 0.2 - \frac{(0.2 - 1.2)1.6167}{1.6167 - 2.7808} = -1.1888.$$

These values will be our next initial values and the same procedure is repeated by using Euler's method with

$$t(0) = u_3 = -1.2383, l(0) = v_3 = -1.1888.$$

Using Euler's formula for i = 0, 1, 2 and 3

$$y_1 = 0.2500, y_2 = 0.4226, y_3 = 0.5335, y_4 = 0.5938.$$

$$z_1 = 0.6904, z_2 = 0.4434, z_3 = 0.2415, z_4 = 0.0733.$$

$$t_1 = -0.9883, t_2 = -0.8073, t_3 = -0.6729, t_4 = -0.5686.$$

$$w_1 = 0.7028, w_2 = 0.4681, w_3 = 0.2637, w_4 = 0.0723.$$

$$l_1 = -0.9388, l_2 = -0.8175, l_3 = -0.7656, l_4 = -0.7497.$$

The difference between the boundary conditions $y'(1) = z(1) = 0$ and $w(1) = 0$ and the calculated values $z_4 = 0.0733, w_4 = 0.0723$ are

$$a(3) = 0.0733 - 0 = 0.0733,$$

$$b(3) = 0.0723 - 0 = 0.0723.$$

We will stop here, as it can be seen that the difference reduces more after each iteration, and we will get more accurate results in the next two iterations.

Main steps for solving the secant method combined with the Euler's method (shooting method for coupled nonlinear equations) in MATLAB: The output is z, the inputs are the function 'func', initial condition x0, step size h and final point of domain xf.

1. For the coupled system (third order and second order) two guesses are chosen randomly, i.e.,
   u1 = 1.3,
   v1 = 1.2,
   y1 = [0 1 u1 1 v1].
2. The domain is
   x = x0:h:xf.
3. The column vector is defined by
   y(:,1) = y1.
4. Using Euler's method
   y(:,i + 1)=y(:,i) + h*func(x(i),y(:,i)).
5. In column form we can write the results as
   t = y'.
6. The last values from the second column and the fourth column are subtracted from the boundary conditions.
   a1 = t(end,2) – 0 (a1= last value of the second column – the first boundary condition),
   b1 = t(end,4) – 0 (b1 = last value of the fourth column – the second boundary condition).
7. Again, different initial guesses (u2 = 0.3, v2 = 0.2) are assumed and the same procedure is repeated.
   a2 = t(end,2) – 0,
   b2 = t(end,4) – 0.
8. A loop is generated for the refinement in initial guesses by the secant method.
9. The initial guesses and functions are replaced with
   s = a1,
   r = a2,
   q = u1,
   u = u2,
   the secant formula for the first equation is
   u3 = u – ((u – q)/(r – s)) * (r).
   For the second equation, initial guesses and functions are replaced with
   c = b1,
   d = b2,
   e = v1,
   v = v2,
   the secant formula for the second equation is
   v3 = v – ((v – e)/(d – c)) * (d).
10. Euler's method is used again with the calculated results of the secant method.
11. The last values are
    t = y',
    w = t(end,2) – 0,
    g = t(end,4) – 0.
12. The results are modified by

   a1 = r,
   a2 = w,
   u1 = u,
   u2 = u3,
   b1 = d,
   b2 = g,
   v1 = v,
   v2 = v3.
   This will refine the values in the secant method and new initial guesses are achieved and
   the guesses are refined until they satisfy the difference between the calculated values
   and the boundary conditions.
13. The second script is used to write the system
    $z = [y(2); y(3); (y(2)^2) - y(1) * y(3); y(5); y(4) * y(2)]$.
14. Using the command window both files are run, and the required result is calculated.

```
% Mfile 1
function [z]=system(func,x0,h,xf)
% z is output
% func, y1, x0,h and xf are the inputs
u1=1.3;
v1=1.2;
y1=[0 1 u1 1 v1];
x=x0:h:xf; % Domain
y(:,1)=y1; % One column
for i=1:length(x)-1
  y(:,i+1)=y(:,i)+h*func(x(i),y(:,i)); % Euler's formula
end
t=y';
a1=t(end,2)-0;
b1=t(end,4)-0;
u2=0.3;
v2=0.2;
y1=[0 1 u2 1 v2];
x=x0:h:xf; % Domain
y(:,1)=y1; % One column
for i=1:length(x)-1
  y(:,i+1)=y(:,i)+h*func(x(i),y(:,i)); % Euler's formula
end
t=y';
a2=t(end,2)-0;
b2=t(end,4)-0;
for numiter=1:6
  s=a1; % Calculated values for first equation
  r=a2;
  q=u1;
  u=u2;
u3=u -((u-q)/(r-s))*(r);% Secant method for first equation
  c=b1; % Calculated values for second equation
  d=b2;
  e=v1;
  v=v2;
v3=v -((v-e)/(d-c))*(d);% Secant method for second equation
for i=1:length(x)-1
y1=[0 1 u3 1 v3];
x=x0:h:xf; % Domain
```

```
y(:,1)=y1; % One column
   y(:,i+1)=y(:,i)+h*func(x(i),y(:,i)) % Euler's formula
end
t=y';
w=t(end,2)-0;
a1=r; % For modification in first equation guesses
a2=w;
u1=u;
u2=u3;
g=t(end,4)-0;
b1=d; % For modification in second equation guesses
b2=g;
v1=v;
v2=v3;
end
z=t;
plot(x,z(:,2))
hold on
plot(x,z(:,1))
end
```

```
% Mfile 2
function [z]=example_system(x,y)
z=[y(2); y(3); (y(2)^2)-y(1)*y(3); y(5); y(4)*y(2)];
end
```

```
>> % Command Window
>> system(@example_system,0,0.25,1)
```

7.3.3   Newton Raphson Method Combined with Euler's Method: For Second Order Nonlinear Ordinary Differential Equations

Suppose a second order nonlinear boundary value problem

$$y''(x,\alpha) = f\left(x, y(x,\alpha), y'(x,\alpha)\right), y(a,\alpha) = A, y(b,\alpha) = B. \tag{7.31}$$

Converting the boundary value problem into an initial value problem

$$y''(x,\alpha) = f\left(x, y(x,\alpha), y'(x,\alpha)\right), y(a,\alpha) = A, y'(a,\alpha) = \alpha. \tag{7.32}$$

Taking the partial derivative of the above mentioned initial value problem with respect to α, we get

$$
\begin{aligned}
\frac{\partial y''(x,\alpha)}{\partial \alpha} &= \frac{\partial f\left(x, y(x,\alpha), y'(x,\alpha)\right)}{\partial x} \frac{\partial x}{\partial \alpha} \\
&+ \frac{\partial f\left(x, y(x,\alpha), y'(x,\alpha)\right)}{\partial y} \frac{\partial y}{\partial \alpha} \\
&+ \frac{\partial f\left(x, y(x,\alpha), y'(x,\alpha)\right)}{\partial y'} \frac{\partial y'}{\partial \alpha}.
\end{aligned}
\tag{7.33}
$$

As x and α are independent, we have $\dfrac{\partial x}{\partial \alpha} = 0$, so the equation becomes

$$\frac{\partial y''(x,\alpha)}{\partial \alpha} = \frac{\partial f\left(x,y(x,\alpha),y'(x,\alpha)\right)}{\partial y}\frac{\partial y}{\partial \alpha}$$
$$+ \frac{\partial f\left(x,y(x,\alpha),y'(x,\alpha)\right)}{\partial y'}\frac{\partial y'}{\partial \alpha}. \tag{7.34}$$

The initial conditions become

$$\frac{\partial y(a,\alpha)}{\partial \alpha} = 0, \frac{\partial y'(a,\alpha)}{\partial \alpha} = 1. \tag{7.35}$$

Denote $\dfrac{\partial y}{\partial \alpha} = q$, the problem becomes

$$q''(x,\alpha) = \frac{\partial f}{\partial y}q(x,\alpha) + \frac{\partial f}{\partial y'}q'(x,\alpha). \tag{7.36}$$

With conditions $q(a,\alpha) = 0$ and $q'(a,\alpha) = 1$.

The system is solved by using any initial value method and the initial guess is modified by using the Newton Raphson method, i.e.,

$$\alpha_{i+1} = \alpha_i - \frac{y(b,\alpha_i) - B}{q(b,\alpha_i)}. \tag{7.37}$$

Example 7.5: Use the shooting method (Euler's and Newton Raphson method) to calculate the solution of the nonlinear boundary value problem

$$y'' = -\frac{(y')^2}{y},$$

$$y(1) = \sqrt{2}, y(3) = 2.$$

Solution: We have

$$f(x,y,y') = -\frac{(y')^2}{y}.$$

The partial derivative with respect to y and y' gives

$$\frac{\partial f}{\partial y} = \frac{(y')^2}{y^2} \quad \text{and} \quad \frac{\partial f}{\partial y'} = \frac{-2y'}{y}.$$

The system becomes

$$y'' = -\frac{(y')^2}{y}, y(1) = \sqrt{2}, y'(1) = u_1$$

$$q'' = \frac{(y')^2}{y^2}q - \frac{2y'}{y}q', q(1) = 0, q'(1) = 1.$$

Convert the system into first order. For this let

$$z = y',$$

so

$$z' = -\frac{(z)^2}{y},$$

and

$$l = q',$$

so

$$l' = \frac{(z)^2}{y^2} q - \frac{2z}{y} l.$$

With initial conditions

$$y(1) = \sqrt{2}, z(1) = u_1 = 1, q(1) = 0, l(1) = 1.$$

We have $h = 0.5, x_0 = 1, y_0 = \sqrt{2}, z_0 = 1, q_0 = 0, l_0 = 1$.
Since $x_0 = 1, x_1 = x_0 + h = 1.5, x_2 = x_0 + 2h = 2, x_3 = x_0 + 3h = 2.5, x_4 = x_0 + 4h = 3$.
Using Euler's formula for $i = 0, 1, 2$ and 3, we have
  $y_1 = 1.9142, y_2 = 2.2374, y_3 = 2.5061, y_4 = 2.7425$.
  $z_1 = 0.6464, z_2 = 0.5373, z_3 = 0.4728, z_4 = 0.4282$.
  $q_1 = 0.5000, q_2 = 0.6464, q_3 = 0.7577, q_4 = 0.8515$.
  $l_1 = 0.2929, l_2 = 0.2225, l_3 = 0.1877, l_4 = 0.1658$.
  There is a difference between the boundary condition which is $y(3) = 2$ and the calculated value $y(3) = 2.7425$. The difference is

$$y(b, u_1) - 2 = 2.7425 - 2 = 0.7425.$$

Now the Newton Raphson method is used to modify the result, i.e.,

$$u_{i+1} = u_i - \frac{y(b, u_i) - B}{q(b, u_i)}.$$

For $i = 1$, we get

$$u_2 = u_1 - \frac{y(b, u_1) - 2}{q(b, u_1)} = 1 - \frac{0.7425}{0.8515} = 0.1281.$$

This value is our next initial value, and the same procedure is repeated by using Euler's method.
Using Euler's formula for $i = 0, 1, 2$ and 3
  $y_1 = 1.4783, y_2 = 1.5394, y_3 = 1.5980, y_4 = 1.6544$.
  $z_1 = 0.1223, z_2 = 0.1172, z_3 = 0.1128, z_4 = 0.1088$.
  $q_1 = 0.5000, q_2 = 0.9547, q_3 = 1.3727, q_4 = 1.7602$.
  $l_1 = 0.9094, l_2 = 0.8359, l_3 = 0.7750, l_4 = 0.7237$.
  Again, the difference between the boundary condition and the calculated value $y(3)$ is

$$y(b, u_2) - 2 = 1.6544 - 2 = -0.3456.$$

Again, using the Newton Raphson method for $i = 2$, we get $u_3 = 0.3244$. Using Euler's formula for $i = 0, 1, 2$ and 3
  $y_1 = 1.5764, \quad y_2 = 1.7200, \quad y_3 = 1.8506, \quad y_4 = 1.9712$.

$z_1 = 0.2872$, $z_2 = 0.2611$, $z_3 = 0.2412$, $z_4 = 0.2255$.
$q_1 = 0.5000$, $q_2 = 0.8853$, $q_3 = 1.2045$, $q_4 = 1.4804$.
$l_1 = 0.7706$, $l_2 = 0.6385$, $l_3 = 0.5518$, $l_4 = 0.4901$.

As we can see, the difference between the boundary condition $y(3) = 2$ and the calculated value $y_3 = 1.9712$ is small. So, we can stop here or can refine the guess by using one more iteration to get a more accurate result.

Main steps for solving the Newton Raphson method combined with Euler's method (shooting method for single second order nonlinear ordinary differential equation) in MATLAB:

1. For second order ordinary differential equations, four systems are generated after converting them into first order form.
2. Euler's method is applied by choosing the initial condition.
3. The final values of the first and third system are used.
4. The last value from the first equation is subtracted from the boundary condition and is denoted by a.
   (a = the last value – the boundary condition).
5. The last value from the third equation is denoted by b.
6. A loop is generated before applying the Newton Raphson method.
7. Apply the Newton Raphson method by using *initial value* – (4)/(5).
8. The result is used in Euler's method and if the final value of the first equation satisfies the boundary condition, then we are done; otherwise, the final value of the first equation and the final value of the third equation are used again in the Newton Raphson method and the procedure is stopped whenever it satisfies the boundary condition. For modified values, this procedure is performed in a loop.
9. The second Mfile is used to represent the equation in first order form.
10. Both first and second Mfiles are run in the command window.

```
% Mfile 1
function [t]=Newton_shooting_single(func,x0,h,xf)
% z is the output
% func, y1, x0,h and xf are the inputs
u(1)=1;
y1=[sqrt(2) u(1) 0 1];
x=x0:h:xf; % Domain
y(:,1)=y1; % One column
for i=1:length(x)-1
y(:,i+1)=y(:,i)+h*func(x(i),y(:,i)); % Euler's formula
end
a=y(1,end)-2;
b=y(3,end);
for numiter=1:15
s=a;
  r=b;
  u=u(1);
  u2=u-(s/r);
for i=1:length(x)-1
y1=[sqrt(2) u2 0 1];
x=x0:h:xf; % Domain
y(:,1)=y1; % One column
y(:,i+1)=y(:,i)+h*func(x(i),y(:,i)); % Euler's formula
end
s=y(1,end)-2;
r=y(3,end);
  u(1)=u2;
```

```
   a=s;
    b=r;
end
t =y';
end
```

```
% Mfile 2
function [z]=Newton_shooting_example(x,y)
z=[y(2); -(y(2)^2)/y(1); y(4);((y(2)^2)/(y(1))^2)*y(3)-2*(y(2)/y(
1))*y(4)];
end
```

```
>> % Command Window
>> Newton_shooting_single(@Newton_shooting_example,1,0.5,3)
```

7.3.4 NEWTON RAPHSON METHOD COMBINED WITH EULER'S METHOD: FOR COUPLED NONLINEAR ORDINARY DIFFERENTIAL EQUATIONS

Consider a third and second order nonlinear boundary value problem

$$y'''(x,\alpha) = f\left(x, y(x,\alpha), y'(x,\alpha), y''\ (x,\alpha), w(x,\alpha), w'(x,\alpha)\right), y(a,\alpha)$$

$$= A, y'(a,\alpha) = B, y(b,\alpha) = C, \tag{7.38}$$

$$w''(x,\alpha) = g\left(x, y(x,\alpha), y'(x,\alpha), y''\ (x,\alpha), w(x,\alpha), w'(x,\alpha)\right), w(a,\alpha)$$

$$= D, w(b,\alpha) = E. \tag{7.39}$$

The initial value problem becomes

$$y'''\ (x,\alpha) = f\left(x, y(x,\alpha), y'(x,\alpha), y''\ (x,\alpha), w(x,\alpha), w'(x,\alpha)\right), y(a,\alpha)$$

$$= A, y'(a,\alpha) = B, y''(a,\alpha) = \alpha, \tag{7.40}$$

$$w''(x,\alpha) = g\left(x, y(x,\alpha), y'(x,\alpha), y''\ (x,\alpha), w(x,\alpha), w'(x,\alpha)\right), w(a,\alpha)$$

$$= D, w'(a,\alpha) = \alpha. \tag{7.41}$$

Taking the partial derivative of both equations with respect to α

$$\frac{\partial y'''(x,\alpha)}{\partial \alpha} = \frac{\partial f}{\partial x}\frac{\partial x}{\partial \alpha} + \frac{\partial f}{\partial y}\frac{\partial y}{\partial \alpha}$$

$$+ \frac{\partial f}{\partial y'}\frac{\partial y'}{\partial \alpha} + \frac{\partial f}{\partial y''}\frac{\partial y''}{\partial \alpha} + \frac{\partial f}{\partial w}\frac{\partial w}{\partial \alpha} + \frac{\partial f}{\partial w'}\frac{\partial w'}{\partial \alpha}, \tag{7.42}$$

$$\frac{\partial w''(x,\alpha)}{\partial \alpha} = \frac{\partial g}{\partial x}\frac{\partial x}{\partial \alpha} + \frac{\partial g}{\partial y}\frac{\partial y}{\partial \alpha}$$

$$+ \frac{\partial g}{\partial y'}\frac{\partial y'}{\partial \alpha} + \frac{\partial g}{\partial y''}\frac{\partial y''}{\partial \alpha} + \frac{\partial g}{\partial w}\frac{\partial w}{\partial \alpha} + \frac{\partial g}{\partial w'}\frac{\partial w'}{\partial \alpha}. \tag{7.43}$$

Put $\dfrac{\partial x}{\partial \alpha} = 0$, the equations become

$$\frac{\partial y'''(x,\alpha)}{\partial \alpha} = \frac{\partial f}{\partial y}\frac{\partial y}{\partial \alpha}$$

$$+ \frac{\partial f}{\partial y'}\frac{\partial y'}{\partial \alpha} + \frac{\partial f}{\partial y''}\frac{\partial y''}{\partial \alpha} + \frac{\partial f}{\partial w}\frac{\partial w}{\partial \alpha} + \frac{\partial f}{\partial w'}\frac{\partial w'}{\partial \alpha}, \tag{7.44}$$

$$\frac{\partial w''(x,\alpha)}{\partial \alpha} = \frac{\partial g}{\partial y}\frac{\partial y}{\partial \alpha}$$

$$+ \frac{\partial g}{\partial y'}\frac{\partial y'}{\partial \alpha} + \frac{\partial g}{\partial y''}\frac{\partial y''}{\partial \alpha} + \frac{\partial g}{\partial w}\frac{\partial w}{\partial \alpha} + \frac{\partial g}{\partial w'}\frac{\partial w'}{\partial \alpha}, \tag{7.45}$$

The initial conditions are

$$\frac{\partial y(a,\alpha)}{\partial \alpha} = 0, \frac{\partial y'(a,\alpha)}{\partial \alpha} = 0, \frac{\partial y''(a,\alpha)}{\partial \alpha} = 1, \tag{7.46}$$

$$\frac{\partial w(a,\alpha)}{\partial \alpha} = 0, \frac{\partial w'(a,\alpha)}{\partial \alpha} = 1. \tag{7.47}$$

Denote $\dfrac{\partial y}{\partial \alpha} = q \ and \ \dfrac{\partial w}{\partial \alpha} = r$, the problem becomes

$$q'''(x,\alpha) = \frac{\partial f}{\partial y}q(x,\alpha) + \frac{\partial f}{\partial y'}q'(x,\alpha) + \frac{\partial f}{\partial y''}q''(x,\alpha) + \frac{\partial f}{\partial w}r(x,\alpha) + \frac{\partial f}{\partial w'}r'(x,\alpha), \tag{7.48}$$

$$r''(x,\alpha) = \frac{\partial g}{\partial y}q(x,\alpha) + \frac{\partial g}{\partial y'}q'(x,\alpha) + \frac{\partial g}{\partial y''}q''(x,\alpha) + \frac{\partial g}{\partial w}r(x,\alpha) + \frac{\partial g}{\partial w'}r'(x,\alpha), \tag{7.49}$$

With conditions $q(a,\alpha) = 0$, $q'(a,\alpha) = 0$ and $q''(a,\alpha) = 1$, $r(a,\alpha) = 0$ and $r'(a,\alpha) = 1$.

The system is solved by using any initial value method and the initial guesses are modified by using Newton's Raphson method, i.e.,

$$\alpha_{i+1} = \alpha_i - \frac{y'(b,\alpha_i) - B}{q'(b,\alpha_i)}, \tag{7.50}$$

$$\beta_{i+1} = \beta_i - \frac{w(b,\beta_i) - E}{r(b,\beta_i)}. \tag{7.51}$$

Example 7.6: Use the shooting method (Euler's and Newton Raphson method) to calculate the solution of the coupled nonlinear boundary value problem

$$y''' + yy'' - y'^2 = 0,$$

$$w'' - y'w = 0,$$

$$y(0) = 0, y'(0) = 1, w(0) = 1, y'(1) = 0, w(1) = 0.$$

Solution: We have

$$f(x, y, y', w, w') = -yy'' + y'^2,$$

$$g(x, y, y', w, w') = y'w.$$

The partial derivatives with respect to y, y', w, w' and y'' give

$$\frac{\partial f}{\partial y} = -y'', \frac{\partial f}{\partial y'} = 2y', \frac{\partial f}{\partial y''} = -y, \frac{\partial g}{\partial y'} = w, \frac{\partial g}{\partial w} = y'.$$

The system becomes

$y''' = -yy'' + y'^2$, $y(0) = 0, y'(0) = 1, y''(0) = u_1,$
$q''' = -y''q + 2y'q' - yq'', q(0) = 0, q'(0) = 0, q''(0) = 1,$
$w'' = y'w$, $w(0) = 1, w'(0) = u_2,$

$$r'' = wq', r(0) = 0, r'(0) = 1.$$

Convert the system into first order. For this let

$y' = z$, $z' = y'' = t$, $t' = -yt + z^2$, $q' = l$, $l' = m$, $m' = -tq + 2zl - ym$, $w' = o$, $o' = zw$, $r' = c$, $c' = wl$.

With initial conditions

$$y(0) = 0, z(0) = 1, t(0) = u_1 = 0.1, q(0) = 0, l(0) = 0, m(0) = 1,$$

$$w(0) = 1, \quad o(0) = u_2 = 0.1, \quad r(0) = 0, \quad c(0) = 1.$$

We have $h = 0.5, x_0 = 0, y_0 = 0, z_0 = 1, t_0 = 0.1, q_0 = 0, l_0 = 0, m_0 = 1, w_0 = 1, o_0 = 0.1, r_0 = 0, c_0 = 1.$
Since $x_0 = 0, x_1 = x_0 + h = 0.5, x_2 = x_0 + 2h = 1.$
Now by applying Euler's method for i = 0 and 1.
$y_1 = 0.5000$, $y_2 = 1.0250.$

$z_1 = 1.0500$, $z_2 = \mathbf{1.3500}.$
$t_1 = 0.6000$, $t_2 = 1.0013.$
$q_1 = 0.0000$, $q_2 = 0.2500.$
$l_1 = 0.5000$, $l_2 = 1.0000.$
$m_1 = 1.0000$, $m_2 = 1.2750.$
$w_1 = 1.0500$, $w_2 = \mathbf{1.3500}.$
$o_1 = 0.6000$, $o_2 = 1.1513.$
$r_1 = 0.5000$, $r_2 = 1.0000.$
$c_1 = 1.0000$, $c_2 = 1.2625.$

The difference between the calculated values z_2, w_2 and the boundary conditions are

$$y'(b, u_1) - 0 = \mathbf{1.3500} - 0 = \mathbf{1.3500},$$

$$w(b, v_1) - 0 = \mathbf{1.3500} - 0 = \mathbf{1.3500}.$$

Now Newton's Raphson method is used to modify the result, i.e.,

$$u_{i+1} = u_i - \frac{y'(b, u_i) - 0}{q'(b, u_i)}, \ v_{i+1} = v_i - \frac{w(b, v_i) - 0}{r(b, v_i)}.$$

For i = 1, we get $u_2 = 0.1 - \dfrac{1.3500}{1} = -1.2500, \ v_2 = 0.1 - \dfrac{1.3500}{1} = -1.2500.$

These values are new initial values, and the same procedure is repeated by using Euler's method.

Using Euler's formula for i = 0 and 1
$y_1 = 0.5000, y_2 = 0.6875.$

$z_1 = 0.3750, \ z_2 = \mathbf{0}.$
$t_1 = -0.7500, \ t_2 = -0.4922.$
$q_1 = 0.0000, \ q_2 = 0.2500.$
$l_1 = 0.5000, \ l_2 = 1.0000.$
$m_1 = 1.0000, \ m_2 = 0.9375.$
$w_1 = 0.3750, \ w_2 = \mathbf{0}.$
$o_1 = 0.7500, \ o_2 = -0.6797.$
$r_1 = 0.5000, \ r_2 = 1.0000.$
$c_1 = 1.0000, \ c_2 = 1.0938.$

As the difference between the boundary conditions and the calculated values (z_2 and w_2) are equal, so we stop here.

Main steps for solving the Newton Raphson method combined with Euler's method (shooting method for systems of equations) in MATLAB: Changes are made in the Newton Raphson method and in the system of equations.

1. In the first order form a system of equations is generated.
2. Two initial guesses are chosen for the system.
   u(1) = 0.1,
   v(1) = 0.1,
   y1 = [0 1 u(1) 0 0 1 1 v(1) 0 1].
3. Ten systems are solved by using Euler's method.
4. The first and second condition of the first equation is given, we choose the third condition as a guess. Similarly for the second equation the first condition is given but the second condition is chosen as the initial guess.
5. Euler's method is implemented, the final values of the second system and the seventh system are subtracted from the given boundary conditions of the first and second equations.
6. The initial guesses are refined by using the Newton Raphson method, the final value of the second system is subtracted from the first boundary condition and the fifth's system final value is used in the Newton Raphson method as a modified initial guess. Similarly, the seventh system final value is subtracted from the second boundary condition with the final value of the ninth system being used in the Newton Raphson method.
7. A loop is applied before applying the Newton Raphson method.
8. These new guesses are used in Euler's method.
9. The guesses are modified by the Newton Raphson method and the procedure is repeated until the given boundary conditions are achieved.
10. The second Mfile is used to write ten systems of equations.
11. The results are displayed in the command window by using both files.

```
% Mfile 1
function [t]=shooting_Newton_systems(func,x0,h,xf)
% z is the output
% func, y1, x0,h and xf are the inputs
u(1)=0.1;
v(1)=0.1;
y1=[0 1 u(1) 0 0 1 1 v(1) 0 1];
x=x0:h:xf; % Domain
y(:,1)=y1; % One column
for i=1:length(x)-1
  y(:,i+1)=y(:,i)+h*func(x(i),y(:,i)); % Euler's formula
end
a=y(2,end)-0;
b=y(5,end);
aa=y(7,end)-0;
bb=y(9,end);
for numiter=1:10
s=a;
  r=b;
  u=u(1);
u2=u-(s/r); % Newton Raphson method for the first equation
  ss=aa;
  rr=bb;
  uu=v(1);
uu2=uu-(ss/rr);% Newton Raphson method for the second equation
for i=1:length(x)-1
y1=[0 1 u2 0 0 1 1 uu2 0 1];
x=x0:h:xf; % Domain
y(:,1)=y1; % One column
  y(:,i+1)=y(:,i)+h*func(x(i),y(:,i)) % Euler's formula
end
s=y(2,end)-0;
r=y(5,end);
ss=y(7,end)-0;
rr=y(9,end);
% Modifying guesses and functions for Newton Raphson method
  u(1)=u2;
  a=s;
  b=r;
  v(1)=uu2;
aa=ss;
  bb=rr;
end
t =y';
end
function [z]=example_Newton_systems(x,y)
z=[y(2);y(3);(y(2)^2)-y(1)*y(3);y(5);y(6);y(3)*y(4)+2*y(2)*y(5)-y
(1)*y(6);y(8);y(2)*y(7);y(10);y(7)*y(5)];
end
```

```
>> % Command Window
>> shooting_Newton_systems(@example_Newton_systems,0,0.1,1)
```

7.4 FINITE DIFFERENCE METHOD FOR LINEAR ORDINARY DIFFERENTIAL EQUATIONS

There is another way of solving the boundary value problem known as the finite difference method. The differentials appearing in the equations are replaced with the approximated results of finite differences at equally spaced mesh points. As a result, the differential equations are transformed into a set of algebraic equations. These sets of equations may be linear or nonlinear. The nonlinear equations are converted into linear algebraic form. These linear equations take the form of a matrix, whose solution is calculated by Gauss elimination, LU decomposition or any iterative method (discussed in Chapter 2).

7.4.1 FINITE DIFFERENCE APPROXIMATION

The finite difference approximation is based on the forward, central and backward Taylor series expansion of $f(x)$ about point x, such that

$$f(x+h) = f(x) + hf'(x) + \frac{h^2}{2!}f''(x) + \frac{h^3}{3!}f'''(x) + \dots \tag{7.52}$$

$$f(x-h) = f(x) - hf'(x) + \frac{h^2}{2!}f''(x) - \frac{h^3}{3!}f'''(x) + \dots \tag{7.53}$$

$$f(x+2h) = f(x) + 2hf'(x) + \frac{(2h)^2}{2!}f''(x) + \frac{(2h)^3}{3!}f'''(x) + \dots \tag{7.54}$$

$$f(x-2h) = f(x) - 2hf'(x) + \frac{(2h)^2}{2!}f''(x) - \frac{(2h)^3}{3!}f'''(x) + \dots \tag{7.55}$$

The sum and difference of the above mentioned equations are

$$f(x+h) + f(x-h) = 2f(x) + h^2 f''(x) + \dots \tag{7.56}$$

$$f(x+h) - f(x-h) = 2hf'(x) + \frac{h^3}{3}f'''(x) + \dots \tag{7.57}$$

$$f(x+2h) + f(x-2h) = 2f(x) + 4h^2 f''(x) + \dots \tag{7.58}$$

$$f(x+2h) - f(x-2h) = 4hf'(x) + \frac{8h^3}{3}f'''(x) + \dots \tag{7.59}$$

In Eqs. (7.56) to (7.59), the sum contains even derivatives while differences contain odd derivatives.

7.4.2 CENTRAL DIFFERENCE APPROXIMATIONS

From Eq. (7.57), we have

$$f'(x) = \frac{f(x+h) - f(x-h)}{2h} - \frac{h^2}{6}f'''(x) + \dots \tag{7.60}$$

Eq. (7.60) can be written as

$$f'(x) = \frac{f(x+h)-f(x-h)}{2h} + O(h^2).$$ (7.61)

Eq. (7.61) is known as the central difference approximation for the first derivative $f'(x)$. The last term $O(h^2)$ is known as truncation error of order 2.

From Eq. (7.56) we have

$$f''(x) = \frac{f(x+h)-2f(x)+f(x-h)}{h^2} + O(h^2).$$ (7.62)

Eq. (7.62) is known as the central difference approximation for the second derivative $f''(x)$.

Higher order derivatives can be obtained from the above mentioned equations, but in this book we restrict ourselves to the first and second order derivatives.

7.4.3 FINITE DIFFERENCE METHOD FOR SECOND ORDER LINEAR ORDINARY DIFFERENTIAL EQUATIONS

Let us consider the second order linear ordinary differential equation

$$y'' + B(x)y' + A(x)y = C(x), \quad a \le x \le b,$$ (7.63)

subject to the boundary conditions

$$y(a) = d_1, y(b) = d_2.$$ (7.64)

Assume a positive integer n and subdivide the interval $a \le x \le b$ into n equal subintervals using n − 1 grid values $x_1 = a+h, x_2 = a+2h, \ldots x_{n-1} = a+(n-1)h$, where h = (b − a)/n.

The first and end values are $x_0 = a$ and $x_n = b$ (as shown in Figure 7.4).

The finite difference approximation for the differential equation (7.63) at $x = x_i$ is

$$\frac{y_{i+1} - 2y_i + y_{i-1}}{h^2} + B(x_i)\frac{y_{i+1} - y_{i-1}}{2h} + A(x_i)y_i = C(x_i).$$ (7.65)

Multiplying Eq. (7.65) by h^2 and rearranging the equation yields

$$y_{i-1}(1 - \frac{B(x_i)h}{2}) + y_i(A(x_i)h^2 - 2) + y_{i+1}(1 + \frac{B(x_i)h}{2}) = C(x_i)h^2, i = 2,3\ldots n-1$$ (7.66)

or

$$b_i y_{i-1} + a_i y_i + c_i y_{i+1} = (x_i)h^2.$$ (7.67)

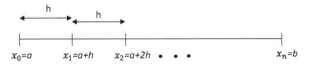

FIGURE 7.4 Grid points for finite difference method.

The boundary conditions become

$$y_0 = d_1, y_n = d_2.$$ (7.68)

Where $b_i = \left(1 - \dfrac{B(x_i)h}{2}\right)$, $a_i = \left(A(x_i)h^2 - 2\right)$ and $c_i = \left(1 + \dfrac{B(x_i)h}{2}\right)$.

The system of Eq. (7.67) in matrix form after using the boundary conditions becomes

$$AY = r,$$ (7.69)

where

$$A = \begin{bmatrix} 1 & 0 & \cdots & \cdots & 0 \\ b_2 & a_2 & c_2 & \cdots & 0 \\ \cdots & \cdots & \cdots & \cdots & \cdots \\ \cdots & \cdots & \cdots & \cdots & \cdots \\ 0 & 0 & b_{n-1} & a_{n-1} & c_{n-1} \\ 0 & 0 & 0 & 1 \end{bmatrix}, \quad Y = \begin{bmatrix} y_1 \\ y_2 \\ y_3 \\ . \\ . \\ y_n \end{bmatrix} \text{ and } r = \begin{bmatrix} d_1 \\ C(x_2)h^2 \\ C(x_3)h^2 \\ . \\ . \\ d_2 \end{bmatrix}.$$

The linear system Eq. (7.69) is tridiagonal and can be solved directly ($Y = A^{-1}r$) or by LU decomposition method, Gauss elimination or any iterative method (discussed briefly in Chapter 2).

Example 7.7: Find the solution of the boundary value problem

$$y'' = y + t,$$

$$y(0) = 0, y(1) = 0.$$

With $h = \dfrac{1}{4} = 0.25$.

Solution: Apply the central difference formula, we get

$$\frac{y_{i+1} - 2y_i + y_{i-1}}{h^2} = y_i + t_i, i = 1, 2, 3.$$

Multiply h^2 on both sides we get

$$y_{i-1} + y_i(-2 - h^2) + y_{i+1} = h^2 t_i.$$

In matrix form we have

$$A = \begin{bmatrix} (-2 - h^2) & 1 & 0 \\ 1 & (-2 - h^2) & 1 \\ 0 & 1 & (-2 - h^2) \end{bmatrix}, \quad X = \begin{bmatrix} x_1 \\ x_2 \\ x_3 \end{bmatrix} \text{ and } r = \begin{bmatrix} h^2 t_1 \\ h^2 t_2 \\ h^2 t_3 \end{bmatrix}.$$

The given tridiagonal system is solved by using Doolittle's method for the tridiagonal system (briefly discussed in Chapter 2).

Where $t = [t_1, t_2, t_3] = [0.2500, \ 0.5000, 0.7500]$. For h = 0.25 we have

$$A = \begin{bmatrix} \left(-2-h^2\right) & 1 & 0 \\ 1 & \left(-2-h^2\right) & 1 \\ 0 & 1 & \left(-2-h^2\right) \end{bmatrix} = \begin{bmatrix} -2.0625 & 1 & 0 \\ 1 & -2.0625 & 1 \\ 0 & 1 & -2.0625 \end{bmatrix}.$$

By using Doolittle's method for tridiagonal matrices

$$\begin{bmatrix} -2.0625 & 1 & 0 \\ 1 & -2.0625 & 1 \\ 0 & 1 & -2.0625 \end{bmatrix} = \begin{bmatrix} \alpha_1 & \gamma_1 & 0 \\ \beta_2\gamma_1 & \beta_2\gamma_1 + \alpha_2 & \gamma_2 \\ 0 & \beta_3\gamma_2 & \beta_3\gamma_2 + \alpha_3 \end{bmatrix},$$

$\gamma_1 = 1, \gamma_2 = 1, \ \alpha_1 = -2.0625, \ \beta_2 = -0.4848, \ \alpha_2 = -1.5777, \ \beta_3 = -0.6339, \ \alpha_3 = -1.4286$.
So

$$U = \begin{bmatrix} -2.0625 & 1 & 0 \\ 0 & -1.5777 & 1 \\ 0 & 0 & -1.4286 \end{bmatrix} \text{ and } L = \begin{bmatrix} 1 & 0 & 0 \\ -0.4848 & 1 & 0 \\ 0 & -0.6339 & 1 \end{bmatrix}.$$

Calculate $LY = R$

$$\begin{bmatrix} 1 & 0 & 0 \\ \beta_2 & 1 & 0 \\ 0 & \beta_3 & 1 \end{bmatrix}\begin{bmatrix} y_1 \\ y_2 \\ y_3 \end{bmatrix} = \begin{bmatrix} r_1 \\ r_2 \\ r_3 \end{bmatrix},$$

this gives
$y_1 = r_1, \ y_2 = r_2 - y_1\beta_2, \ y_3 = r_3 - y_2\beta_3,$
where

$$r = \begin{bmatrix} h^2 t_1 \\ h^2 t_2 \\ h^2 t_3 \end{bmatrix} = \begin{bmatrix} 0.0156 \\ 0.0312 \\ 0.0469 \end{bmatrix}.$$

Hence,
$y_1 = 0.0156, \ y_2 = r_2 - y_1\beta_2 = 0.0312 - (0.0156)(-0.4848) = 0.0388,$
$y_3 = r_3 - y_2\beta_3 = 0.0469 - (0.0388)(-0.6339) = 0.0715.$
Calculating $x_1, \ x_2$ and x_3

$$x_3 = \frac{y_3}{\alpha_3} = 0.0715 / (-1.4286) = -0.0500,$$

$$x_2 = \left(0.0388 - (2)(-0.0349)\right) / (-1.5777) = -0.0563,$$

$$x_1 = (y_1 - \gamma_1 x_2) / \alpha_1$$

$$= \left(0.0156 - (1)(-0.0563)\right) / (-2.0625)$$

$$= -0.0349.$$

Main steps for solving the linear ordinary differential equation by using the finite difference method in MATLAB: The procedure is the same as solving the tridiagonal banded matrix (as discussed in Chapter 2).

1) Define the step size h, number of iteration n, and initial point t(1) i.e.,
   n = 3, h = 1/4, t(1) = 1/4.
   The domain is defined in the loop from 2:n and the domain is
   t(j) = t(j − 1) + h.
2) Define the diagonal, lower diagonal and upper diagonal entries in the loop.
3) The remaining steps are the same as were briefly discussed in Chapter 2.

```
% Mfile
function x = Finite_difference_Ex1
n=5;
h=1/4;
t(1)=0;
for j=2:n
t(j)=t(j-1)+h;
end
a(1)=1;
for j=2:n-1
  a(j)=-2-h^2; % Diagonal entries
end
a(n)=1;
for j=2:n-1
  b(j)=1; % Lower diagonal entries
end
b(n)=0;
c(1)=0;
for j=2:n-1
  c(j)=1; % Upper diagonal entries
end
for j=1:n-1
gama(j)=c(j);
end
alpha(1)=a(1);
for j=2:n
  beta(j)=b(j)/alpha(j-1)
  alpha(j)=a(j)-beta(j)*gama(j-1)
end
r(1)=0;
for j=2:n-1
  r(j)=(h^2)*t(j);
end
r(n)=0;
y(1)=r(1);
for j=2:n
  y(j)=r(j)-y(j-1)*beta(j);
end
x(n)=y(n)/alpha(n);
for j=n-1:-1:1  % Back substitution UX=Y
  x(j)=(y(j)-gama(j)*x(j+1))/alpha(j);
end
```

7.4.4 FINITE DIFFERENCE METHOD FOR THE SYSTEM OF LINEAR ORDINARY DIFFERENTIAL EQUATIONS

Let us consider the system of linear differential equations

$$y'' + B(x)y' + A(x)y = g, \quad a \leq x \leq b, \tag{7.70}$$

$$g'' + C(x)g' + D(x)g = y, \tag{7.71}$$

with boundary conditions

$$y(a) = u_1, y(b) = u_3,$$

$$g(a) = u_2, g(b) = u_4. \tag{7.72}$$

Subdivide the interval $a \leq x \leq b$ into n equal subintervals using n − 1 grid values $x_1 = a + h, x_2 = a + 2h, \ldots x_{n-1} = a + (n-1)h$, where h = (b − a)/n.

The finite difference approximation for the differential systems Eqs. (7.70 to 7.71) at $x = x_i$ is

$$\frac{y_{i+1} - 2y_i + y_{i-1}}{h^2} + B(x_i)\frac{y_{i+1} - y_{i-1}}{2h} + A(x_i)y_i = g_i, \tag{7.73}$$

$$\frac{g_{i+1} - 2g_i + g_{i-1}}{h^2} + C(x_i)\frac{g_{i+1} - g_{i-1}}{2h} + D(x_i)g_i = y_i, \tag{7.74}$$

Multiplying Eqs. (7.73) and (7.74) by h^2 and rearranging the equation gives

$$y_{i-1}(1 - \frac{B(x_i)h}{2}) + y_i\left(A(x_i)h^2 - 2\right) - h^2 g_i + y_{i+1}(1 + \frac{B(x_i)h}{2}) = 0, \tag{7.75}$$

$$g_{i-1}(1 - \frac{C(x_i)h}{2}) + g_i\left(D(x_i)h^2 - 2\right) - h^2 y_i + g_{i+1}(1 + \frac{C(x_i)h}{2}) = 0, \quad i = 1, 2..n \tag{7.76}$$

or

$$b_i y_{i-1} + a_i y_i - g_i h^2 + c_i y_{i+1} = 0, \tag{7.77}$$

$$d_i g_{i-1} + e_i g_i - h^2 y_i + f_i g_{i+1} = 0, \tag{7.78}$$

the boundary conditions become

$$y_0 = u_1, g_0 = u_2, y_n = u_3, g_n = u_4. \tag{7.79}$$

Where $b_i = (1 - \frac{B(x_i)h}{2})$, $a_i = \left(A(x_i)h^2 - 2\right)$, $c_i = (1 + \frac{B(x_i)h}{2})$, $d_i = (1 - \frac{C(x_i)h}{2})$, $e_i = \left(D(x_i)h^2 - 2\right)$

and $f_i = (1 + \frac{C(x_i)h}{2})$.

System along with boundary conditions Eq. (7.79) can be written as

$$
\begin{bmatrix}
\begin{bmatrix} 1 & 0 \\ 0 & 1 \end{bmatrix} & \begin{bmatrix} 0 & 0 \\ 0 & 0 \end{bmatrix} & \cdots & \cdots & \begin{bmatrix} 0 & 0 \\ 0 & 0 \end{bmatrix} \\
\begin{bmatrix} b_1 & 0 \\ 0 & d_1 \end{bmatrix} & \begin{bmatrix} a_1 & -h^2 \\ -h^2 & e_1 \end{bmatrix} & \begin{bmatrix} c_1 & 0 \\ 0 & f_1 \end{bmatrix} & \cdots & \begin{bmatrix} 0 & 0 \\ 0 & 0 \end{bmatrix} \\
& \cdots & \cdots & \cdots & & \cdots \\
& \cdots & \cdots & \cdots & \\
\begin{bmatrix} 0 & 0 \\ 0 & 0 \end{bmatrix} & \cdots & \begin{bmatrix} b_{n-1} & 0 \\ 0 & d_{n-1} \end{bmatrix} & \begin{bmatrix} a_{n-1} & h^2 \\ -h^2 & e_{n-1} \end{bmatrix} & \begin{bmatrix} c_{n-1} & 0 \\ 0 & f_{n-1} \end{bmatrix} \\
\begin{bmatrix} 0 & 0 \\ 0 & 0 \end{bmatrix} & \cdots & \cdots & \cdots & \begin{bmatrix} 1 & 0 \\ 0 & 1 \end{bmatrix}
\end{bmatrix}
\begin{bmatrix}
\begin{bmatrix} y_0 \\ g_0 \end{bmatrix} \\
\begin{bmatrix} y_1 \\ g_1 \end{bmatrix} \\
- \\ - \\ - \\ - \\
\begin{bmatrix} y_{n-1} \\ g_{n-1} \end{bmatrix} \\
\begin{bmatrix} y_n \\ g_n \end{bmatrix}
\end{bmatrix}
$$

$$(7.80)$$

$$
=
\begin{bmatrix}
\begin{bmatrix} u_1 \\ u_2 \end{bmatrix} \\
\begin{bmatrix} 0 \\ 0 \end{bmatrix} \\
- \\ - \\ - \\ - \\
\begin{bmatrix} 0 \\ 0 \end{bmatrix} \\
\begin{bmatrix} u_3 \\ u_4 \end{bmatrix}
\end{bmatrix}.
$$

The system is solved by using Crout's method. MATLAB is used to solve this system as the system is large, so it is difficult to solve manually. For n = 3 or 5 the given end boundary condition (target) is not achievable. For n > 100 we will get the desired result.

Example 7.8: Find the solution of the boundary value problem

$$y'' = g, g'' = y,$$

with boundary conditions

$$y(0) = 1, y(1) = 0, g(0) = 1, g(1) = 0,$$

and $h = \dfrac{1}{4} = 0.25$.

Solution: The finite difference approximation for the differential system at $x = x_i$ is

$$\frac{y_{i+1} - 2y_i + y_{i-1}}{h^2} = g_i, \quad \frac{g_{i+1} - 2g_i + g_{i-1}}{h^2} = y_i,$$

multiplying above equation by h^2 and rearranging gives

$$y_{i-1} - 2y_i + y_{i+1} - h^2 g_i = 0,$$

$$g_{i-1} - 2g_i + g_{i+1} - h^2 y_i = 0.$$

The boundary conditions become

$$y_0 = 1, g_0 = 1, y_n = 0, g_n = 0.$$

By using the system along with the boundary conditions, we have the following form

$$
\begin{bmatrix}
\begin{bmatrix} 1 & 0 \\ 0 & 1 \end{bmatrix} & \begin{bmatrix} 0 & 0 \\ 0 & 0 \end{bmatrix} & \cdots & \cdots & \begin{bmatrix} 0 & 0 \\ 0 & 0 \end{bmatrix} & \\
\begin{bmatrix} 1 & 0 \\ 0 & 1 \end{bmatrix} & \begin{bmatrix} -2 & -h^2 \\ -h^2 & -2 \end{bmatrix} & \begin{bmatrix} 1 & 0 \\ 0 & 1 \end{bmatrix} & \cdots & \begin{bmatrix} 0 & 0 \\ 0 & 0 \end{bmatrix} & \\
 & \cdots & \cdots & \cdots & & \cdots \\
 & \cdots & \cdots & \cdots & & \\
\begin{bmatrix} 0 & 0 \\ 0 & 0 \end{bmatrix} & \cdots & \begin{bmatrix} 1 & 0 \\ 0 & 1 \end{bmatrix} & \begin{bmatrix} -2 & h^2 \\ -h^2 & -2 \end{bmatrix} & \begin{bmatrix} 1 & 0 \\ 0 & 1 \end{bmatrix} & \\
\begin{bmatrix} 0 & 0 \\ 0 & 0 \end{bmatrix} & \cdots & \cdots & \cdots & \begin{bmatrix} 1 & 0 \\ 0 & 1 \end{bmatrix} &
\end{bmatrix}
\begin{bmatrix}
\begin{bmatrix} y_0 \\ g_0 \end{bmatrix} \\
\begin{bmatrix} y_1 \\ g_1 \end{bmatrix} \\
- \\ - \\ - \\ - \\
\begin{bmatrix} y_{n-1} \\ g_{n-1} \end{bmatrix} \\
\begin{bmatrix} y_n \\ g_n \end{bmatrix}
\end{bmatrix}
$$

$$
=
\begin{bmatrix}
\begin{bmatrix} 1 \\ 1 \end{bmatrix} \\
\begin{bmatrix} 0 \\ 0 \end{bmatrix} \\
- \\ - \\ - \\ - \\
\begin{bmatrix} 0 \\ 0 \end{bmatrix} \\
\begin{bmatrix} 0 \\ 0 \end{bmatrix}
\end{bmatrix}.
$$

The system is solved by using Crout's method in MATLAB.

Main steps for solving the system of linear ordinary differential equations by using the finite difference method in MATLAB: Crout's method is used to solve the system of linear differential equations.

1) For the satisfaction of boundary conditions, n must be greater than 100.
2) Define the step size h, the initial point xx(1) and the domain in the loop.
3) The entries which will appear in the matrix will be defined.
4) The diagonal, lower diagonal and upper diagonal entries are defined in the loop.
5) Define the right side as

r1(1) = 1, r2(1) = 1.
6) The remaining entries are zero. So, we generate a loop in which the remaining entries
   are mentioned as zero.
   r1(j) = 0, r2(j) = 0 for j = 1:n.
7) Apply Crout's method for the block banded tridiagonal system.
8) After getting the solution, we will break the solution by using
   k(:,j) = x{j}(1,1),
   z(:,j) = x{j}(2,1).
9) This will break the block into solutions separately in y and z form.

```
% Mfile
function Block_Band_Crouts_tridiagnol_system
xx(1)=0;
n=200;
h=0.1;
for i=2:n
  xx(i)=xx(i-1)+h;
end
a{1}=[1 0;0 1];
c{1}=[0 0;0 0];
for j=2:n-1
  a{j}=[-2 -h^2;-h^2 -2];
end
a{n}=[1 0;0 1];
for j=2:n-1
  b{j}=[1 0;0 1];
end
b{n}=[0 0;0 0];
for j=2:n-1
  c{j}=[1 0;0 1];
end
c{n}=[0 0;0 0];
r1(1)=1;
  r2(1)=1;
for j=2:n
  r1(j)=0;
  r2(j)=0;
end
gamma{1}=inv(a{1})*c{1};
for j=2:n
  a{j}=a{j}-(b{j}*gamma{j-1});
  gamma{j}=inv(a{j})*c{j};
end
for j=1:n
  rr{j}=[r1(j);r2(j)];
end
y{1}=inv(a{1})*rr{1};
for j=2:n
  y{j}=inv(a{j})*(rr{j}-b{j}*y{j-1});
end
x{n}=y{n};
for j=n-1:-1:1
  x{j}=y{j}-(gamma{j})*x{j+1};
end
for j=n:-1:1
```

```
    k(:,j)=x{j}(1,1);
    z(:,j)=x{j}(2,1);
end
k=k';
z=z';
plot(xx,z)
```

7.5   FINITE DIFFERENCE METHOD FOR THE NONLINEAR ORDINARY DIFFERENTIAL EQUATIONS

In this section, we will briefly discuss the finite difference scheme for the nonlinear boundary value problems. In general, the nonlinear boundary value problems are more difficult than the linear ones. Consider a nonlinear boundary value problem of the form

$$y'' = f(x, y, y'), \quad y(\alpha) = y_\alpha, \quad y(\beta) = y_\beta. \tag{7.81}$$

Assume that the function $f(x, y, y')$ has continuous derivatives which satisfy

$$0 < S_* \le f_y(x, y, y') \le S^* \text{ and } |f_y(x, y, y')| \le M^*. \tag{7.82}$$

For some constants S_*, S^* and M^*. Like in the previous section, the central difference formulas are applied to get the nonlinear algebraic form. The nonlinear form is linearized by any iterative method. In this chapter, Newton's iterative method is used to linearize the nonlinear algebraic equations.

7.5.1   FINITEE DIFFERENCE METHOD FOR THE SECOND ORDER NONLINEAR ORDINARY DIFFERENTIAL EQUATIONS

Consider the second order nonlinear ordinary differential equation

$$y'' + B(x)y' + A(x)y^2 = D(x), \quad a \le x \le b, \tag{7.83}$$

with boundary conditions

$$y(a) = d_1, y(b) = d_2. \tag{7.84}$$

Subdivide the interval $a \le x \le b$ into n equal subintervals using n − 1 grid values $x_0 = a$, $x_1 = a + h$, $x_2 = a + 2h, \ldots x_{n-1} = a + (n-1)h$, $x_n = b$. where h = (b-a)/n.

Apply finite difference approximation to the differential equation (7.83) at $x = x_i$

$$\frac{y_{i+1} - 2y_i + y_{i-1}}{h^2} + B(x_i)\frac{y_{i+1} - y_{i-1}}{2h} + A(x_i)y_i^2 = D(x_i), \tag{7.85}$$

multiplying Eq. (7.85) by h^2 gives

$$y_{i+1} - 2y_i + y_{i-1} + \frac{B(x_i)h}{2}(y_{i+1} - y_{i-1}) + h^2 A(x_i)y_i^2 = h^2 D(x_i). \tag{7.86}$$

Eq. (7.86) cannot be solved directly due to the nonlinear term. So, firstly, we linearize this equation and then use the previous method to solve the tridiagonal matrix.

Newton's Method

To linearize Eq. (7.86) we use Newton's method, introducing the following iterate

$$y_i^{k+1} = y_i^k + \delta y_i^k. \tag{7.87}$$

Substituting Eq. (7.87) into Eq. (7.86) and dropping the quadratic and higher order terms in δ, moreover shifting the terms which are free from δ on the right-hand side, we get

$$\delta y_{i+1} - 2\delta y_i + \delta y_{i-1} + \frac{B(x_i)h}{2}(\delta y_{i+1} - \delta y_{i-1}) + 2h^2 A(x_i) y_i \delta y_i = h^2 D(x_i) - y_{i+1} + 2y_i - y_{i-1}$$

$$-\frac{B(x_i)h}{2}(y_{i+1} - y_{i-1}) - h^2 A(x_i) y_i^2 \tag{7.88}$$

or

$$\delta y_{i-1}\left(1 - \frac{B(x_i)h}{2}\right) + \delta y_i\left(-2 + 2h^2 A(x_i) y_i\right) + \delta y_{i+1}\left(1 + \frac{B(x_i)h}{2}\right)$$

$$= h^2 D(x_i) - y_{i+1} + 2y_i - y_{i-1} - \frac{B(x_i)h}{2}(y_{i+1} - y_{i-1}) - h^2 A(x_i) y_i^2. \tag{7.89}$$

The above Eq. (7.89) can be written as

$$b_i \delta y_{i-1} + a_i \delta y_i + c_i \delta y_{i+1} = r_i, \quad i = 2, 3 .. n - 1 \tag{7.90}$$

with boundary conditions

$$\delta y_1 = 0 \text{ and } \delta y_n = 0. \tag{7.91}$$

Where

$$b_i = 1 - \frac{B(x_i)h}{2}, a_i = -2 + 2h^2 A(x_i) y_i, c_i = 1 + \frac{B(x_i)h}{2} \quad \text{and} \quad r_i = h^2 D(x_i) - y_{i+1} + 2y_i - y_{i-1}$$

$$-\frac{B(x_i)h}{2}(y_{i+1} - y_{i-1}) - h^2 A(x_i) y_i^2.$$

Eq. (7.90) is the linear algebraic equation, the coefficients b_i, a_i, c_i and r_i are known quantities. The question arises what type of known quantities? So, the answer is that we must choose an initial guess (known quantities) which satisfies the boundary conditions ($y(a) = d_1, y(b) = d_2$).

Eq. (7.90) along with boundary conditions Eq. (7.91) can be written in matrix form

$$\begin{bmatrix} 1 & 0 & \cdots & \cdots & 0 \\ b_2 & a_2 & c_2 & \cdots & 0 \\ \cdots & \cdots & \cdots & \cdots & \cdots \\ \cdots & \cdots & \cdots & \cdots & \cdots \\ 0 & 0 & b_{n-1} & a_{n-1} & c_{n-1} \\ 0 & 0 & 0 & 0 & 1 \end{bmatrix} \begin{bmatrix} \delta y_1 \\ \delta y_2 \\ \delta y_3 \\ . \\ . \\ \delta y_n \end{bmatrix} = \begin{bmatrix} 0 \\ r_2 \\ r_3 \\ . \\ . \\ 0 \end{bmatrix}. \tag{7.92}$$

The tridiagonal system Eq. (7.92) is solved by the LU decomposition method (already discussed in Chapter 2). The solution is not calculated yet, the solutions $\delta y_1, \delta y_2 ....\delta y_n$ are added to the initial guess to get the desired result, i.e.,

$$y_i^{k+1} = y_i^k + \delta y_i^k. \tag{7.93}$$

Example 7.9: Find the solution of the boundary value problem

$$y'' = 1.5y^2,$$

with boundary conditions

$$y(0) = 4, y(1) = 1.$$

and $h = 0.01$.
Solution: Apply the central difference formula, we get

$$\frac{y_{i+1} - 2y_i + y_{i-1}}{h^2} = 1.5y_i^2,$$

multiply h^2 on both sides we get

$$y_{i-1} - 2y_i + y_{i+1} = 1.5h^2 y_i^2.$$

The nonlinear equation is linearized by using Newton's method, introducing the iterate

$$y_i^{k+1} = y_i^k + \delta y_i^k.$$

Substituting this relation into a nonlinear equation and dropping the quadratic and higher order terms in δ, moreover shifting the terms which are free from δ on the right-hand side, we get

$$\delta y_{i-1}(1) + \delta y_i(-2 - 3h^2 y_i) + \delta y_{i+1}(1) = -y_{i+1} + 2y_i - y_{i-1} - 1.5h^2 y_i^2.$$

The above equation can be written as

$$b_i \delta y_{i-1} + a_i \delta y_i + c_i \delta y_{i+1} = r_i, \quad i = 2, 3..n-1$$

with boundary conditions

$$\delta y_1 = 0 \text{ and } \delta y_n = 0.$$

Where
$b_i = 1, a_i = -2 - 3h^2 y_i, c_i = 1$ and $r_i = -y_{i+1} + 2y_i - y_{i-1} - 1.5h^2 y_i^2$.
The linear equation along with boundary conditions can be written in matrix form

$$\begin{bmatrix} 1 & 0 & \cdots & \cdots & 0 \\ b_2 & a_2 & c_2 & \cdots & 0 \\ \cdots & \cdots & \cdots & \cdots & \cdots \\ \cdots & \cdots & \cdots & \cdots & \cdots \\ 0 & 0 & b_{n-1} & a_{n-1} & c_{n-1} \\ 0 & & 0 & 0 & 1 \end{bmatrix} \begin{bmatrix} \delta y_1 \\ \delta y_2 \\ \delta y_3 \\ . \\ . \\ \delta y_n \end{bmatrix} = \begin{bmatrix} 0 \\ r_2 \\ r_3 \\ . \\ . \\ 0 \end{bmatrix}.$$

The system is solved by the LU decomposition method. The calculated values of $\delta y_1, \delta y_2 \ldots \delta y_n$ are substituted in (7.93) to get the result. In Eq. (7.93) y_i^k is the initial guess which satisfies the boundary conditions.

Main steps for solving the second order nonlinear ordinary differential equation by using the finite difference method in MATLAB:

1. Define the step size h, number of iteration n, initial point t(1) and the domain.
   n = 101, h = 0.01, t(1) = 0, t(j) = t(j-1) + h.
2. Next define an initial guess which satisfies the boundary conditions. For current problem
   aa(j,k) = 4 – t(j) – t(j) ^ 2 – t(j)^3, where j = 1 to n.
   Note: The value of k is 1.
3. Implement the diagonal, lower diagonal and upper diagonal entries in the loop. Note that the initial guess is used as a known quantity. The values of r_i are implemented in a similar manner.
4. Apply the LU decomposition method to calculate the values of δy_i.
5. At the end both the initial guess y_i and δy_i are added to get the result.

```
% Mfile
function xx=Finite_difference_nonlinear_second_order
n=101;
h=0.01;
t(1)=0;
for j=2:n
t(j)=t(j-1)+h;
end
k=1;
for j=1:n
aa(j,k)=4-t(j)-t(j)^2-t(j)^3;  % Guess
end
 for j=2:n-1
    a(1,k)=1;
a(j,k)=-2-3*h^2*aa(j,k);
   end
  a(n,k)=1;
for j=2:n
  b(j,k)=1; % Lower diagonal entries
end
c(1,k)=0;
for j=2:n-1
  c(j,k)=1; % Upper diagonal entries
end
for j=1:n-1
gama(j,k)=c(j,k);
end
alpha(1,k)=a(1,k);
for j=2:n
  beta(j,k)=b(j,k)/alpha(j-1,k);
 alpha(j,k)=a(j,k)-beta(j,k)*gama(j-1,k);
end
r(1,k)=0;
for j=2:n-1
  r(j,k+1)=1.5*h^2*(aa(j,k))^2-aa(j+1,k)-aa(j-1,k)+2*aa(j,k);
end
```

```
r(n,k) =0;
y(1,k)=r(1,k);
for j=2:n
 y(j,k)=r(j,k)-y(j-1,k)*beta(j,k);
end
x(n,k)=y(n,k)/alpha(n,k);
for j=n-1:-1:1  % Back substitution UX=Y
 x(j,k)=(y(j,k)-gama(j,k)*x(j+1,k))/alpha(j,k);
end
for j=1:n
xx(j,k)= aa(j,k)+x(j,k);
end
```

7.5.2 FINITE DIFFERENCE METHOD FOR THE SYSTEM OF NONLINEAR ORDINARY DIFFERENTIAL EQUATIONS

Let us consider the system of nonlinear ordinary differential equations

$$y'' + B(x)y' + A(x)y^2 = g, \quad a \le x \le b, \tag{7.94}$$

$$g'' + C(x)g' + D(x)g^2 = y, \tag{7.95}$$

with boundary conditions

$$y(a) = u_1, y(b) = u_3,$$

$$g(a) = u_2, g(b) = u_4. \tag{7.96}$$

Subdivide the interval $a \le x \le b$ into n equal subintervals using n – 1 grid values $x_1 = a + h, x_2 = a + 2h, \ldots x_{n-1} = a + (n-1)h$, where h = (b – a)/n.

The finite difference approximation for the ordinary differential Eqs. (7.94) and (7.95) at $x = x_i$ are

$$\frac{y_{i+1} - 2y_i + y_{i-1}}{h^2} + B(x_i)\frac{y_{i+1} - y_{i-1}}{2h} + A(x_i)y_i^2 = g_i, \tag{7.97}$$

$$\frac{g_{i+1} - 2g_i + g_{i-1}}{h^2} + C(x_i)\frac{g_{i+1} - g_{i-1}}{2h} + D(x_i)g_i^2 = y_i. \tag{7.98}$$

Multiplying Eqs. (7.97) and (7.98) by h^2 gives

$$y_{i+1} - 2y_i + y_{i-1} + \frac{B(x_i)h}{2}(y_{i+1} - y_{i-1}) + h^2 A(x_i)y_i^2 = h^2 g_i, \tag{7.99}$$

$$g_{i+1} - 2g_i + g_{i-1} + \frac{C(x_i)h}{2}(g_{i+1} - g_{i-1}) + h^2 D(x_i)g_i^2 = h^2 y_i. \tag{7.100}$$

Newton's Method

   Linearize Eqs. (7.99) and (7.100) by using Newton's method

$$y_i^{k+1} = y_i^k + \delta y_i^k, \tag{7.101}$$

$$g_i^{k+1} = g_i^k + \delta g_i^k. \tag{7.102}$$

Substituting Eqs. (7.101) and (7.102) into Eqs. (7.99) and (7.100) and dropping the quadratic and higher order terms in δ, moreover shifting the terms which are free from δ on the right-hand side, we get

$$\begin{aligned}
&\delta y_{i+1} - 2\delta y_i + \delta y_{i-1} + \frac{B(x_i)h}{2}(\delta y_{i+1} - \delta y_{i-1}) \\
&+2h^2 A(x_i) y_i \delta y_i - h^2 \delta g_i = -y_{i+1} + 2y_i - y_{i-1} \\
&-\frac{B(x_i)h}{2}(y_{i+1} - y_{i-1}) \\
&-h^2 A(x_i) y_i^2 + h^2 g_i,
\end{aligned} \tag{7.103}$$

$$\begin{aligned}
&\delta g_{i+1} - 2\delta g_i + \delta g_{i-1} + \frac{C(x_i)h}{2}(\delta g_{i+1} - \delta g_{i-1}) \\
&+2h^2 D(x_i) g_i \delta g_i - h^2 \delta y_i \\
&= -g_{i+1} + 2g_i - g_{i-1} - \frac{C(x_i)h}{2}(g_{i+1} - g_{i-1}) \\
&-h^2 D(x_i) g_i^2 + h^2 y_i,
\end{aligned} \tag{7.104}$$

with boundary conditions reduce to

$$\delta y(a) = 0, \delta y(b) = 0,$$

$$\delta g(a) = 0, \delta g(b) = 0. \tag{7.105}$$

Assume

$$r_1(x_i) = -y_{i+1} + 2y_i - y_{i-1} - \frac{B(x_i)h}{2}(y_{i+1} - y_{i-1}) - h^2 A(x_i) y_i^2 + h^2 g_i,$$

$$r_2(x_i) = -g_{i+1} + 2g_i - g_{i-1} - \frac{C(x_i)h}{2}(g_{i+1} - g_{i-1}) - h^2 D(x_i) g_i^2 + h^2 y_i.$$

   Eqs. (7.103) and (7.104) can be written as

$$\delta y_{i-1}\left(1 - \frac{B(x_i)h}{2}\right) + \delta y_i\left(-2 + 2h^2 A(x_i) y_i\right) + \delta y_{i+1}\left(1 + \frac{B(x_i)h}{2}\right) + \delta g_i(-h^2) = r_1(x_i), \tag{7.106}$$

$$\delta g_{i-1}\left(1 - \frac{C(x_i)h}{2}\right) + \delta g_i\left(-2 + 2h^2 D(x_i)g_i\right) + \delta g_{i+1}\left(1 + \frac{C(x_i)h}{2}\right) + \delta y_i(-h^2) = r_2(x_i). \quad (7.107)$$

Eqs. (7.106) and (7.107) can be written as

$$(7.108)$$

The linear system Eq. (7.108) is a banded block tridiagonal and can be solved by the LU decomposition method. After solving this system, the calculated values are substituted in Eqs. (7.101) and (7.102) to get the required result.

Example 7.10: Find the solution of the boundary value problem

$$y'' + y' + y^2 = g, \quad 0 \le x \le 20,$$

$$g'' + g' + g^2 = y,$$

with boundary conditions

$$y(0) = 1, \quad y(20) = 0, \quad g(0) = 1, \quad g(20) = 0.$$

and $h = 0.1$.

Solution: The finite difference approximations for the ordinary differential equations at $x = x_i$ are

$$\frac{y_{i+1} - 2y_i + y_{i-1}}{h^2} + \frac{y_{i+1} - y_{i-1}}{2h} + y_i^2 = g_i,$$

$$\frac{g_{i+1} - 2g_i + g_{i-1}}{h^2} + \frac{g_{i+1} - g_{i-1}}{2h} + g_i^2 = y_i.$$

Multiplying the equations by h^2 gives

$$y_{i+1} - 2y_i + y_{i-1} + \frac{h}{2}(y_{i+1} - y_{i-1}) + h^2 y_i^2 = h^2 g_i,$$

$$g_{i+1} - 2g_i + g_{i-1} + \frac{h}{2}(g_{i+1} - g_{i-1}) + h^2 g_i^2 = h^2 y_i.$$

Linearize the equations by using Newton's method
$$y_i^{k+1} = y_i^k + \delta y_i^k, \ g_i^{k+1} = g_i^k + \delta g_i^k.$$
Substituting this relation into equations and dropping the quadratic and higher order terms in δ, moreover shifting the terms which are free from δ on the right-hand side, we get

$$\delta y_{i+1} - 2\delta y_i + \delta y_{i-1} + \frac{h}{2}(\delta y_{i+1} - \delta y_{i-1})$$

$$+2h^2 y_i \delta y_i - h^2 \delta g_i = -y_{i+1} + 2y_i - y_{i-1}$$

$$-\frac{h}{2}(y_{i+1} - y_{i-1}) - h^2 y_i^2 + h^2 g_i,$$

$$\delta g_{i+1} - 2\delta g_i + \delta g_{i-1} + \frac{h}{2}(\delta g_{i+1} - \delta g_{i-1})$$

$$+2h^2 g_i \delta g_i - h^2 \delta y_i = -g_{i+1} + 2g_i - g_{i-1}$$

$$-\frac{h}{2}(g_{i+1} - g_{i-1})$$

$$-h^2 g_i^2 + h^2 y_i,$$

with boundary conditions reduce to

$$\delta y(0) = 0, \ \delta y(20) = 0, \ \delta g(0) = 0, \ \delta g(20) = 0.$$

Assume

$$r_1(x_i) = -y_{i+1} + 2y_i - y_{i-1} - \frac{h}{2}(y_{i+1} - y_{i-1}) - h^2 y_i^2 + h^2 g_i,$$

$$r_2(x_i) = -g_{i+1} + 2g_i - g_{i-1} - \frac{h}{2}(g_{i+1} - g_{i-1}) - h^2 g_i^2 + h^2 y_i.$$

Equations can be written as

$$\delta y_{i-1}\left(1 - \frac{h}{2}\right) + \delta y_i\left(-2 + 2h^2 y_i\right) + \delta y_{i+1}\left(1 + \frac{h}{2}\right) + \delta g_i(-h^2) = r_1(x_i),$$

$$\delta g_{i-1}\left(1 - \frac{h}{2}\right) + \delta g_i\left(-2 + 2h^2 g_i\right) + \delta g_{i+1}\left(1 + \frac{h}{2}\right) + \delta y_i(-h^2) = r_2(x_i).$$

The equations can be written in matrix form as

$$
\begin{bmatrix}
\begin{bmatrix} 1 & 0 \\ 0 & 1 \end{bmatrix} & \begin{bmatrix} 0 & 0 \\ 0 & 0 \end{bmatrix} & \cdots & & \cdots & \begin{bmatrix} 0 & 0 \\ 0 & 0 \end{bmatrix} \\[2ex]
\begin{bmatrix} 1-\dfrac{h}{2} & 0 \\ 0 & 1-\dfrac{h}{2} \end{bmatrix} & \begin{bmatrix} -2+2h^2 y_2 & -h^2 \\ -h^2 & -2+2h^2 g_2 \end{bmatrix} & \begin{bmatrix} 1+\dfrac{h}{2} & 0 \\ 0 & 1+\dfrac{h}{2} \end{bmatrix} & \cdots & & \begin{bmatrix} 0 & 0 \\ 0 & 0 \end{bmatrix} \\[2ex]
& \cdots & \cdots & & \cdots & \\
& \cdots & \cdots & & \cdots & \\
\begin{bmatrix} 0 & 0 \\ 0 & 0 \end{bmatrix} & \cdots & \begin{bmatrix} 1-\dfrac{h}{2} & 0 \\ 0 & 1-\dfrac{h}{2} \end{bmatrix} & \begin{bmatrix} -2+2h^2 y_{n-1} & -h^2 \\ -h^2 & -2+2h^2 g_{n-1} \end{bmatrix} & \begin{bmatrix} 1+\dfrac{h}{2} & 0 \\ 0 & 1+\dfrac{h}{2} \end{bmatrix} \\[2ex]
\begin{bmatrix} 0 & 0 \\ 0 & 0 \end{bmatrix} & \cdots & & \begin{bmatrix} 0 & 0 \\ 0 & 0 \end{bmatrix} & \begin{bmatrix} 1 & 0 \\ 0 & 1 \end{bmatrix}
\end{bmatrix}
$$

$$
\begin{bmatrix}
\begin{bmatrix} \delta y_1 \\ \delta g_1 \end{bmatrix} \\
\begin{bmatrix} \delta y_2 \\ \delta g_2 \end{bmatrix} \\
- \\ - \\ - \\
\begin{bmatrix} \delta y_{n-1} \\ \delta g_{n-1} \end{bmatrix} \\
\begin{bmatrix} \delta y_n \\ \delta g_n \end{bmatrix}
\end{bmatrix}
=
\begin{bmatrix}
\begin{bmatrix} 0 \\ 0 \end{bmatrix} \\
\begin{bmatrix} r_1(x_2) \\ r_2(x_2) \end{bmatrix} \\
- \\ - \\ - \\
\begin{bmatrix} r_1(x_{n-1}) \\ r_2(x_{n-1}) \end{bmatrix} \\
\begin{bmatrix} 0 \\ 0 \end{bmatrix}
\end{bmatrix}.
$$

The system is solved by using the LU decomposition method and the calculated values are substituted into Eqs. (7.101) and (7.102) to get the result.

Main steps for solving the system of nonlinear ordinary differential equations by using the finite difference method in MATLAB:

1) The step size h, number of iteration n, initial point t(1) and the domain are defined
   n = 200, h = 0.1, t(1) = 0, t(j) = t(j – 1) + h, where j = 2 to n.
2) The initial guesses for the current problem are
   aa(j,k) = e^{-t}, bb(j,k) = e^{-t}.
3) Write the block entries in the diagonal, lower diagonal and upper diagonal.
4) The LU decomposition method is applied to calculate the values of δy_i and δg_i.
5) At the end both the initial guesses (y_i, z_i) and δy_i and δz_i are added to get the result.

```
% Mfile
function yy=Finite_difference_nonlinear_system
n=200;
h=0.1;
t(1)=0;
for j=2:n
t(j)=t(j-1)+h;
end
```

```
k=1;
for j=1:n
aa(j,k)=exp(-t(j)); % Guess
bb(j,k)=exp(-t(j));
end
a{1,k}=[1 0;0 1];
  for j=2:n-1
a{j,k}=[-2+2*h^2*aa(j,k) -h^2;-h^2 -2+2*h^2*bb(j,k)];
  end
  a{n,k}=[1 0;0 1];
for j=2:n-1
  b{j,k}=[-1-h/2 0;0 1-h/2]; % Lower diagonal entries
end
b{n,k}=[0 0;0 0];
c{1,k}=[0 0;0 0];
for j=2:n
  c{j,k}=[1+h/2 0;0 1+h/2]; % Upper diagonal entries
end
for j=1:n-1
gama{j,k}=c{j,k};
end
alpha{1,k}=a{1,k};
for j=2:n
  beta{j,k}=inv(alpha{j-1,k})*b{j,k};
  alpha{j,k}=a{j,k}-beta{j,k}*gama{j-1,k};
end
r1(1,k)=0;
r2(1,k)=0;
for j=2:n-1
  r1(j,k)=-aa(j+1,k)+2*aa(j,k)-aa(j-1,k)-(h/2)*(aa(j+1,k)-aa(j-1,k)
)-h^2*((aa(j,k))^2)+(h^2)*bb(j,k);
  r2(j,k)=-bb(j+1,k)+2*bb(j,k)-bb(j-1,k)-(h/2)*(bb(j+1,k)-bb(j-1,k)
)-h^2*((bb(j,k))^2)+(h^2)*aa(j,k);
end
r1(n,k)= 0;
r2(n,k)=0;
for j=1:n
  rr{j,k}=[r1(j,k);r2(j,k)];
end
gamma{1,k}=inv(a{1,k})*c{1,k};
for j=2:n
  a{j,k}=a{j,k}-(b{j,k}*gamma{j-1,k});
  gamma{j,k}=inv(a{j,k})*c{j,k};
end
y{1,k}=inv(a{1,k})*rr{1,k};
for j=2:n
  y{j,k}=inv(a{j,k})*(rr{j,k}-b{j}*y{j-1,k});
end
x{n,k}=y{n,k};
for j=n-1:-1:1
  x{j,k}=y{j,k}-(gamma{j,k})*x{j+1,k};
end
for j=n:-1:1
  m(j,k)=x{j,k}(1,1);
```

```
   z(j,k)=x{j,k}(2,1);
 end
 for j=1:n
 xx(j,k)= aa(j,k)+m(j,k);
 yy(j,k)= bb(j,k)+z(j,k);
 end
```

Problem Set 7

1. Use the MATLAB code of the linear shooting method to calculate the solutions of the given problems
   (i) $xy'' - y' - x^5 = 0$ with $y(1) = 0.5, y(2) = 4$.

   (ii) $y'' + 9y = 0$ with $y(0) = 0, y(\pi/6) = 1$.
   (iii) $6x^2 y'' + xy' + y = 0$ with $y(1) = 2, y(64) = 12$.

2. Consider a beam freely hinged at ends $x = 0$ and $x = L$, with a uniform load m and tension T. The deflection $y(x)$ is described by the ordinary differential equation

$$y'' - \frac{T}{EI} y - \frac{mx(x-L)}{2EI} = 0,$$

   with boundary conditions (see Figure 7.5)

$$y(0) = 0, y(L) = 0.$$

   Calculate the solution of the problem.
3. Use shooting code to solve the system of equations $y'' = y + y', g'' = g + g'$, with boundary conditions y(0) = 0, g(0) = 0, y(1) = 0 and g(1) = 0.
4. With the help of MATLAB, calculate the solutions of the following problem

$$y'''' + \frac{4}{x} y^3 = 0,$$

   with boundary conditions $y(0) = 0, y'(0) = 0, y''(1) = 0$ and $y'''(1) = 0$.
5. Use MATLAB to calculate the solution of the following problem

$$y'' + y' - y^2 = 0,$$

   with boundary conditions $y(0) = 1$ and $y(1) = 2$.
6. Calculate the solution of the following third order nonlinear boundary value problem by using secant and Euler's method

FIGURE 7.5   Hinged beam.

$$y''' + yy'' - y'^3 = 0,$$

with boundary conditions $y(0) = 0, y'(0) = 1$ and $y(5) = 0$.

7. Solve the following system by using secant and Euler's method

$$f''' + f^2 + 2f + 1 = 0,$$

$$g'' + \sin(x)gf = 0,$$

with boundary conditions $f(0) = 0, f'(0) = 0, f(3) = 0, g(0) = 1$ and $g(3) = 0$.

8. Calculate the solution of the problem by using secant and Euler's method

$$f''' + 2ff' + f = 0,$$

$$g'' + fg' = 0,$$

$$s'' + sg' = 0,$$

with boundary conditions $f(0) = 0, f'(0) = 1, f(10) = 0, g(0) = 1, g(10) = 0, s(0) = 1$ and $s(10) = 0$.

9. Apply Newton's method and Euler's method to calculate the solution of the problem

$$f'' = \frac{1}{8}\left(32 + 2x^3 - ff'\right),$$

with boundary conditions $f(1) = 17$ and $f(3) = 14.3333$.

10. Use Newton's method and Euler's method to find the solution of the problem

$$f'' = f^3 - ff',$$

with boundary conditions $f(1) = 0.5$ and $f(3) = 0.33$.

11. Consider the Van der Pol equation

$$y'' - \mu\left(y^2 - 1\right)y' + y = 0,$$

assume $\mu = 0.5$ with

boundary conditions $y(0) = 0$ *and* $y(2) = 1$.

Calculate the solution by using Newton's method and Euler's method.

12. Compute the solution of the problem by using Newton's method and Euler's method.

$$f'' = -f'^2 - f + \ln(x)$$

$$f(1) = 0 \quad f(2) = \ln(2)$$

13. Use MATLAB to calculate the solution of the problem

$$y'' = y + t(t - 4),$$

by using a finite difference method, the boundary conditions are $y(0) = 0$ and $y(4) = 0$.

14. By using MATLAB, calculate the solution of the following problem

$$y'' - e^y = 0,$$

by using a finite difference method, the boundary conditions are $y(0) = 1$ and $y(1) = 0$.

15. Use MATLAB to calculate the solution of the problem

$$y'' + ty'^2 + t^2 y = 0,$$

by using a finite difference method, the boundary conditions are $y(0) = 1$ and $y(2) = -2$.

16. Consider a ball hit so that it lands 300 feet from the home plate after three seconds. Assume that air resistance acts only against the horizontal component of the flight and is proportional to the horizontal velocity, the ball's motion is

$$f'' + cf' = 0,$$

and

$$g'' + g = 0,$$

here c is a drag coefficient which is taken as 0.5. The initial position is $f(0) = 0, g(0) = 3$ as the landing time is $t_f = 3$ so the final points are $f(3) = 300, g(3) = 0$. Calculate the solution of the problem.

Appendix

CHAPTER 2

Solution 1: Using Cramer's rule code

```
  (i)   [X]=cramer([2 1 1;3 3 1;1 3 2],[1 3 12])
 (ii)   [X]=cramer([1 2 1 2;1 8 1 3;2 6 1 9;2 8 6 2],[1 2 11 9])
(iii)   [X]=cramer([3 8 2 8;2 11 3 6;2 2 9 0;-1 9 16 -2],[11 -2 1 -9])
```

Solution 2: Using Gauss elimination code

```
  (i)   [X]=gauss_elimination([1 9 2;1 2 2;1 1 9],[2 1 2])
 (ii)   [X]=gauss_elimination([1 6 2 3;3 3 2 1;2 1 3 1; 5 3 2 1],[1 2 2
        4])
(iii)   [X]=gauss_elimination([1 2 1 9;2 2 1 9;1 7 2 1; 2 1 8 1],[1 3 1 9]
```

Solution 3: By using Doolittle's code and for V=20 we get

```
       doolittle2([3 -5 2 0;-2 5 0 2;0 1 -3 2;3 0 10 -28],[0 -100 0 0])
```

for V=40 we have

```
doolittle2([3 -5 2 0;-2 5 0 2;0 1 -3 2;3 0 10 -28],[0 -200 0 0])
```

by using Crout's code and for V=20 we get

```
Crouts([3 -5 2 0;-2 5 0 2;0 1 -3 2;3 0 10 -28],[0 -100 0 0])
```

for V=40 we get

```
Crouts([3 -5 2 0;-2 5 0 2;0 1 -3 2;3 0 10 -28],[0 -200 0 0])
```

by using Choleski's code for V=20 we get

```
choleski_decomposition([3 -5 2 0;-2 5 0 2;0 1 -3 2;3 0 10 -28],[0 -100 0
0])
```

for V=40 we get

```
choleski_decomposition([3 -5 2 0;-2 5 0 2;0 1 -3 2;3 0 10 -28],[0 -200 0
0])
```

Solution 4: (i)

```
 doolittle2([-1 1 3;1 9 4;2 3 2],[12 11 9])
 Crouts ([-1 1 3;1 9 4;2 3 2],[12 11 9])
choleski_decomposition([-1 1 3;1 9 4;2 3 2],[12 11 9])
(ii) doolittle2([4 2 3 31 2 1 8;7 2 1 2;5 -2 6 2],[2 1 0 4])
 Crouts ([4 2 3 31 2 1 8;7 2 1 2;5 -2 6 2],[2 1 0 4])
choleski_decomposition ([4 2 3 31 2 1 8;7 2 1 2;5 -2 6 2],[2 1 0 4])
```

```
  (iii)doolittle2([1 -2 2 2;1 -2 2 3;1 2 1 -1;1 -1 3 -1],[2 2 3 2])
  Crouts ([1 -2 2 2;1 -2 2 3;1 2 1 -1;1 -1 3 -1],[2 2 3 2])
  choleski_decomposition ([1 -2 2 2;1 -2 2 3;1 2 1 -1;1 -1 3 -1],[2 2
3 2])
```

Solution 5: Using Mfile

```
% Mfile
function [X]=doolittle21(A) % Output X and inputs A and b
X=zeros(n,1); % X have n rows and one column
y=zeros(n,1); % y have n rows and one column
u=zeros(n,n);
l=zeros(n,n);
for j=1:n
  u(j,j)=A(j,j)-l(j,1:j-1)*u(1:j-1,j);% Daigonal matrix
  l(j,j)=1;
  for i=j+1:n
    l(i,j)=(A(i,j)-l(i,1:j-1)*u(1:j-1,j))/u(j,j)
% Lower diagonal matrix
     u(j,i)=A(j,i)-l(j,1:j-1)*u(1:j-1,i) % Upper
%diagonal matrix
   end
end
% Command Window
doolittle21([-4 1 0;-2 5 0 2;1 -4 0;0 1 4])
```

Solution 6:

```
  Crouts([1 2 0; 1 4 1;0 1 2],[2 3 -1])
choleski_decomposition ([1 2 0; 1 4 1;0 1 2],[2 3 -1])
```

Solution 7: Replace a, b c and r values in the codes mentioned in page number 45 and 49.

Solution 8: Replace a, b c and r values in Banded topics for Doolittle's and Crout's methods.

Solution 9: By using Gauss elimination, Doolittle's or Crout's methods the matrix is solved.

Solution 10: Just the diagonal, lower diagonal and upper diagonal entries are replaced in block banded code of Crout's method.

Solution 11: Just the diagonal, lower diagonal and upper diagonal entries are replaced in block banded code of Crout's method.

Solution 12: By using Gauss Jacobi code and in command window

```
 [X]=Gauss_Jacobi([7 -2 3 0;1 -9 1 1;2 0 10 2;1 -1 1 6],[16 12 10 4],[0 0
0 0])
```

Solution 13: By using Gauss Jacobi code and in command window

```
[X]=Gauss_Jacobi([3 1 1;-2 4 1;-1 2 -6],[4 1 3],[0 0 0])
```

Solution 14: By using Gauss Seidel code and in command window

```
  Gauss_Seidel([4 0 1;1 4 1;-1 0 4],[5 3 10],[0 0 0])
```

Solution 15: By using Gauss Seidel code and in command window

```
Gauss_Seidel([1 -1 2 -1;2 1 -2 -2;-1 2 -4 1;3 0 0 0 -2],[-1 -2 1
-3],[0 0 0 0])
```

Solution 16: By using conjugate code and in command window

```
conj_grad([5 -1 3;4 7 -4;6 -3 2],[3 2 9],[0 0 0])
```

Solution 17: By using conjugate code and in command window

```
conj_grad([3 0 -1;0 4 2;-1 -3 5],[4 10 -10],[0 0 0])
```

Solution 18: By using conjugate code and in command window

```
conj_grad([1.2 0.45 (0.35&0.45) ;0.89 2.59 -0.33 -0.22;0.72 0.77 4.02
-0.88;...
 0.12 0.56 0.67 3.38],[2.61 -11.78 -15.03 11.38],[0 0 0 0])
```

CHAPTER 3

Solution 1: By using Newton forward difference code and in command window

```
Newton_forward_differnce([20 25 30 35 40],[0.342 0.422 0.5 0.573
0.643],34)
```

Solution 2: By using Newton froward difference code and in command window

```
Newton_forward_differnce([0.10 0.15 0.20 0.25 0.30],[0.100 0.151 0.203
0.255 0.309],0.27)
```

Solution 3: By using Newton froward and backward difference codes and in command window

```
Newton_forward_differnce([0.20 0.22 0.24 0.26 0.28 0.30],[1.659 1.669
1.691 1.702 1.713],0.23)
Newton_backward_differnce([0.20 0.22 0.24 0.26 0.28 0.30],[1.659 1.669
1.691 1.702 1.713],0.29)
```

Solution 4: By using Newton backward difference code and in command window

```
Newton_backward_differnce([0 3 6],[1.225 0.905 0.652],5.9)
```

Solution 5: Use 't' in 'syms' command and remaining procedure remains same.

Solution 6: By using the Newton divided difference code and in command window

```
Newton_divided_differnce([1 2 3 4 5 6],[14.6 19.6 30.6 53.6 94.6
159.6],3.4)
```

Solution 7: Firstly, substituting the values of x in the given function to get equal numbers of outputs i.e.,

$$f(-2) = 2(-2)^3+4(-2)^2+2(-2)+1=-3,$$

the remaining values are calculated in a similar way, we will get

X	-2	-1	0	1	2	3	4
f(X)	-3	1	1	9	37	97	201

Now by using these values in the Newton divided difference code and using "syms" command for "t" i.e.,

```
% Mfile
function [p,A]= Newton_divided_differnce(x,y)
% Inputs and outputs are defined.
% x and y are given data
% t is the value which is to be calculated
n=length(x);
A=zeros(n,n); % Consider a n*n matrix
A(:,1)=y';
for j=2:n % The remaining table is calculated
 for i=j:n
  A(i,j)=(A(i,j-1)-A(i-1,j-1))/(x(i)-x(i-j+1));
 end
end
xt=1; % For creating a sequence
s=0; % Adding a sequence
y0=A(1,1); % First entry of y
syms t
for j=1:n-1
xt=xt*(t-x(j)); % Sequence created
s=s+A(j+1,j+1)*xt; % Addition of a sequence
end
p=y0+s; % Formula
```

Solution 8: Put the values of x in the given function, we will get the equal number of inputs and outputs then use the code of Lagrange interpolation method. This procedure can be performed in command window i.e.,

```
% Command Window
for x=1:3
y=1./x
end
```

The results will be

X	1	2	3
f(X)	1	0.5	0.33

```
% Mfile
function [p]= Lagrange_interpolation(x,y)
% Inputs and outputs are defined
% x and y are given data
% xx is the value which is to be calculated
```

```
n=length(x);
t=0; % Addition of sequence
syms xx
for i=1:n
p0=y(i); % Given data
for j=1:n
if i~=j % By formula
   p0=p0*(xx-x(j))/(x(i)-x(j));
end
end
t=t+p0; % Addition of sequence
end
p=t;
```

Solution 9: First calculate the values of the functions i.e.,

(i)

X	0	0.2	0.6
$f(X)=\cos(4X)e^{3X}$	1	1.2695	-4.4610

Now using Lagrange interpolating code and in command window write

```
Lagrange_interpolation([0 0.2 0.6],[1 1.2695 -4.4610])
```

(ii)

X	2	2.3	3
$f(X)=ln(X)$	0.6931	0.8329	1.0986

Using Lagrange interpolating code and in command window write

```
Lagrange_interpolation([2 2.3 3],[0.6931 0.8329 1.0986])
```

(iii)

X	0	0.2	0.6
$f(X)=\cos(X)+2$	3	2.9801	2.8253

Now using Lagrange interpolating code and in command window write

```
Lagrange_interpolation([0 0.2 0.6],[3 2.9801 2.8253])
```

Solution 10: By using Neville's method code and in command window write

```
Nevilles([16.93 17.65 18.42 18.76],[8.1 8.3 8.6 8.8],8.7)
```

Solution 11: (i) Table is constructed

X		0	2	7	9
$f(x) = 2^x$		1	4	128	512

Now develop a Neville's code with few changes

```
% Mfile
function [A]= Nevilles(y,x)
% Inputs and output are defined
% x and y are given data
% yy is the value which is to be calculated
n=length(y);
A=zeros(n,n); % Consider an n*n matrix
syms yy
A(:,1)=y';% As the given values of y is a column vector
for j=2:n
 for i=j:n
A(i,j)=((yy-x(i-j+1))*A(i,j-1)-(yy-x(i))*A(i-1,j-1))/(x(i)-x(i-j+1));
 end
end
% Command Window
Nevilles([0 2 7 9],[1 4 128 512])
```

(ii) Table is constructed

X		0	1	3	7
$f(X) = \sqrt{X}$		0	1	1.7321	2.6458

Now develop a Neville's code

```
% Mfile
function [A]= Nevilles(y,x)
% Inputs and output are defined.
% x and y are given data
% yy is the value which is to be calculated
n=length(y);
A=zeros(n,n); % Consider a n*n matrix
syms yy
A(:,1)=y';% As the given values of y is a column vector
for j=2:n
 for i=j:n
A(i,j)=((yy-x(i-j+1))*A(i,j-1)-(yy-x(i))*A(i-1,j-1))/(x(i)-x(i-j+1));
 end
end
% Command Window
Nevilles([0 1 3 7],[0 1 1.7321 2.6458])
```

(iii) Table is constructed

X	0	2	4	6
$f(X) = sin(2X)$	0	-0.7568	0.9894	-0.5366

Now develop a Neville's code

```
% Mfile
function [A]= Nevilles(y,x)
% Inputs and output are defined.
% x and y are given data
% yy is the value which is to be calculated
n=length(y);
A=zeros(n,n); % Consider a n*n matrix
syms yy
A(:,1)=y';% As the given values of y is a column vector
for j=2:n
 for i=j:n
A(i,j)=((yy-x(i-j+1))*A(i,j-1)-(yy-x(i))*A(i-1,j-1))/(x(i)-x(i-j+1));
 end
end
% Command Window
Nevilles([0 2 4 6],[0 -0.7568 0.9894 -0.5366])
```

Solution 12: The cubic spline code is

```
% Mfile
function [ss]= cubic_spline(b,x,f)
% Inputs and outputs are defined
% x and f are given data
% xx is the value which is to be calculated
n=length(b);
b=b'; % b is a column vector
f=f'; % f is a column vector
A=zeros(n,n);
A(1,1)=1; % First entry of matrix is 1
A(n,n)=1; % Last entry of matrix is 1
b(1)=0; % First entry of b is zero (Natural Spline)
b(n)=0; % Last entry of b is zero (Natural Spline)
function h=h(x,i) % Entries of matrix depends on h
  h=x(i+1)-x(i);
end
for i=2:n-1
% entries are
 A(i,i-1)=h(x,i-1); % Lower diagonal entries
 A(i,i)=2*(h(x,i-1)+h(x,i)); % Diagonal entries
 A(i,i+1)=h(x,i); % Upper diagonal entries
b(i)=((3/h(x,i))*(f(i+1)-f(i))-(3/h(x,i-1))*(f(i)-f(i-1))); % Column
end
c=A\b;
syms xx
for i=1:n-1
```

```
bb(i)=((f(i+1)-f(i))/h(x,i))-(h(x,i)/3)*(2*c(i)+c(i+1));
dd(i)=(c(i+1)-c(i))/(3*h(x,i));
% Finally, all values are calculated from polynomial
ss(i)=f(i)+bb(i)*(xx-x(i))+c(i)*((xx-x(i))^2)+dd(i)*((xx-x(i))^3);
end
end
% Command Window
cubic_spline([0 0 0],[0 1 2],[0 2 1.1])
```

Solution 13: The cubic spline code is used, and command window is used for required output.

```
cubic_spline([0 0 0 0 0],[1.1 2.1 3.1 4.1 5.1],[13.1 15.1 12.1 9.1 13.1],
2.4)
```

Solution 14: The cubic spline code is used, and command window is used for required output.
```
cubic_spline([0 0 0 0],[2 200 2000 20000],[13.90 0.80 0.40 0.43], 10)
```

CHAPTER 4

Solution 1: An Mfile is generated by some name i.e.,

```
% Mfile
function y=abc12(x)
y=-1-x+x^3;
end
```

This file along with bisection Mfile is used in the command window to get the result i.e.,

```
>> % Command Window
>> [x,numiter]=Bisection(@abc12,0,1,6)
```

Solution 2: An Mfile is generated by some name i.e.,

```
% Mfile
function y=abc123(x)
y=2*x-sqrt(1+cos(x));
end
```

This file along with bisection Mfile is used in the command window to get the result i.e.,

```
>> % Command Window
>> [x,numiter]=Bisection(@abc123,0,1,6)
```

Solution 3: An Mfile is generated by some name i.e.,

```
% Mfile
function y=abc1234(x)
y=x^3+2*x^2-3*x-3;
end
```

This file along with bisection Mfile is used in the command window to get the result i.e.,

```
>> % Command Window
>> [x,numiter]=Bisection(@abc1234,0,1,6)
```

Solution 4: An Mfile is generated by some name i.e.,

```
% Mfile
function y=abc12345(x)
y=exp(-x)-x^3-x;
end
```

This file along with regula falsi Mfile is used in the command window to get the result i.e.,

```
>> % Command Window
>> [x,numiter]=False_Position(@abc12345,0,1,5)
```

Solution 5: An Mfile is generated by some name i.e.,

```
% Mfile
function y=abc12346(x)
y=exp(-x)*sin(x)+x+5;
end
```

This file along with regula falsi Mfile is used in the command window to get the result i.e.,

```
>> % Command Window
>> [x,numiter]=False_Position(@abc123456,2.9,3.9,9)
```

Solution 6: An Mfile is generated by some name i.e.,

```
% Mfile
function y=abc123467(x)
y=exp(-x)-x;
end
```

This file along with regula falsi Mfile is used in the command window to get the result i.e.,

```
>> % Command Window
>> [x,numiter]=False_Position(@abc1234567,0,5,12)
```

Solution 7: For Newton Raphson method two more Mfiles are generated, one for the function and second one for the derivative of the function i.e.,

  (i)

```
% Mfile
function y=ast1(x)
y=4-x-3*x^3+x^5;
end
% Mfile
function y=dast1(x)
y=-1-9*x^2+5*x^4;
end
```

These two files along with Newton Raphson code is used to calculate the root in command window.

  (ii)

```
% Mfile
function y=ast12(x)
```

```
y=x*sin(x)+exp(x);
end
% Mfile
function y=dast12(x)
y=sin(x)+x*cos(x)+exp(x);
end
```

These two files along with Newton Raphson code is used to calculate the root in command window.
  (iii)

```
% Mfile
function y=ast123(x)
y=x^3-x^2+exp(x);
end
% Mfile
function y=dast123(x)
y=3*x^2-2*x+exp(x);
end
```

These two files along with Newton Raphson code is used to calculate the root in command window.

Solution 8: Solving the given equation yields

$0.0001x^3 - 4.6x^2 - 0.0003x + 0.0002 = 0$.

The root of the equation is calculated by Newton Raphson method.

Solution 9: Substitute the values and solve the equation by bisection method.

Solution 10:

```
% Mfile
function [x,numiter]=sts12(x)
for numiter=1:12
dx=-(-2-4*x-3*x^2+2*x^3+x^4)/(-4-6*x+6*x^2+4*x^3);
x=x+dx;
end
end
```

Solution 11:

```
% Mfile 1
x1=0.1;
x2=1.2;
% x1 and x2 are the values
% between which the root lies
s=x1;
r=x2;
q=0.1*s^4-3.8*s^3+1.2*s^2+1.43*s+0.21;
u=0.1*r^4-3.8*r^3+1.2*r^2+1.43*r+0.21;
u3=r -((r-s)/(u-q))*(u); % Secant formula
x1=r; % Refined value
x2=u3;
end
c=u3;
end
```

Solution 12: An Mfile is generated by some name i.e.,

```
% Mfile
function y=awer(x)
y= [(x(1)^2)+(x(2)^2)-1;(x(1)^2)-x(2)];
end
```

This file along with the Newton Raphson method for the system of equations is used to calculate the result in command window.

Solution 13: An Mfile is generated by some name i.e.,

```
% Mfile
function y=awer(x)
y= [x(2)-x(1);5*x(1)-x(2)-x(1)*x(3);x(1)*x(2)-16*x(3)];
end
```

This file along with the Newton Raphson method for the system of equations is used to calculate the result in command window.

CHAPTER 5

Solution 1: (a) The two Mfiles are

```
% Mfile
function T=Single_trapezoidal(func,a,b)
% T is the output
x=a; % Replace first value by x
h=b-a;
s=func(a);
m=func(b);
T=(h/2)*(s+m); % Final formula
% Mfile
 function y=rst1(x)
 y=1/(1+x+x^2);
 end
% Command Window
Single_trapezoidal(@rst1,0,2)
```

(b) An Mfile by the name "rst1" is used in the general code as mentioned in page 111.

```
% Command Window
trapezoidal(@rst1,0,2,1)
trapezoidal(@rst1,0,2,2)
```

(c)

```
% Mfile
function S=Sympsons_One_third(func,a,b)
% S is the output
s=func(a); % First value of function
m=func(b); % Final value of function
h=(b-a)/2; % Intervals
```

```
r=func(h);
S=(h/3)*(s+4*r+m);
% Command Window
Sympsons_One_third( (@rst1,0,2)
```

(d) An Mfile by the name "rst1" is used in the general code as mentioned in page 114.

```
% Command Window
S=Sympsons_One_third(@rst1,0,2,2)
```

(e) An Mfile by the name "rst1" is used in the general code as mentioned in page 117.

```
% Command Window
S=Sympsons_three_eight(@rst1,0,2,2)
```

Solution 2: Solutions are calculated in a similar way as calculated in 1, the difference is that n and the function are different i.e.,

```
% Mfile
 function y=rst2(x)
 y=exp(sin(x));
 end
```

Solution 3:

```
% Mfile
function T=trapezoidal1(func,a,b,n)
x=a; % Replace first value by x
s=func(a); % First value of the function
m=func(b); % Final value of the function
h=(b-a)/n; % Interval
t=0; % For sequence to be added
for i=1:n-1
 x=x+h; % x in each interval
 r=2*func(x); % Remaining values of the
 %function rather than first and last ones
 t=t+r; % Addition of all values
end
T=(d/2*b)*(h/2)*(s+m+t); % Final formula
% Mfile
 function y=rst2(x)
d=2;
b=0.01;
t=100;
y=(exp(-d^2/4*b)/(t-x)*sqrt(4*pi*b*(t-x))*(20*sin(2*pi*x/8766)+15);
 end
% Command Window
trapezoidal1(@rst2,0,100,10)
```

Solution 4: Use the codes mentioned in the chapter along with the Mfile i.e.,

```
% Mfile
 function y=rse(x)
y=x/(1+x^2);
 end
```

```
% Command Window
T=trapezoidal(@rse,0,2,10)
% Command Window
S=Sympsons_One_third(@rse,0,2,10)
% Command Window
S=Sympsons_three_eight(@rse,0,2,10)
```

Solution 5: The table values are implemented in a) composite trapezoidal rule, (b) composite Simpson's 1/3 rule and (c) Simpson's 3/8 rule to calculate the integral.

Solution 6: The table values are implemented in a) composite trapezoidal rule, (b) composite Simpson's 1/3 rule and (c) Simpson's 3/8 rule to calculate the integral.

Solution 7: Use the codes mentioned in the chapter along with this Mfile i.e.,

```
% Mfile
 function y=rse2(x)
y=1/(2+cos(5*pi*x));
 end
% Command Window
T=trapezoidal(@rse,0,1,10)
% Command Window
S=Sympsons_One_third(@rse,0,1,10)
% Command Window
S=Sympsons_three_eight(@rse,0,1,10)
```

Solution 8: Use the code mentioned in page 121 along with this Mfile i.e.,

```
% Mfile
 function y=rse22(x)
y=sin(x^2*cos(exp(-x)))*exp(-x);
 end
% Command window
Rombergs(@rse22,0,3,4)
```

Solution 9: Use the code mentioned in page 121 along with this Mfile i.e.,

```
% Mfile
 function y=rse222(x)
y=cos(3*sin(2*x)+3*cos(3*x)+2*cos(2*x)+3*sin(x)+cos(x));
 end
% Command window
Rombergs(@rse222,0,pi,pi/6)
```

Solution 10: Use the code mentioned in page 121 along with this Mfile i.e.,

```
% Mfile
 function y=rse2222(x)
y=2*x^5-4*x^3-x+2;
 end
% Command window
>> Rombergs(@rse2222,-1,3,6)
```

Solution 11: Use the codes mentioned in page 124 and 127 along with this Mfile i.e.,

```
% Mfile
function y=rty(x)
y=(2*x^2+(3/x))^2;
end
```

Command Window is used for results

```
% Command Window
adaptive(@rty,1.1,2.1)
Gausss_quadrature(@rty,1.1,2.1,2)
```

Solution 12: Use the codes mentioned in page 124 and 127 along with this Mfile i.e.,

```
% Mfile
function y=rty1(x)
y=-0.04*x^4+0.872*x^3-4.123*x^2+5.21*x+3;
end
```

Command window is used for results

```
% Command Window
adaptive(@rty1,1,9)
Gausss_quadrature(@rty1,1,9,2)
```

Solution 13: Use the codes mentioned in page 124 and 127 along with this Mfile i.e.,

```
% Mfile
function y=rty2(x)
y=(cos(x))^2/sqrt(4-(sin(x))^2);
end
```

Note: As the values of f_0 and b are not given so "syms" command is used.
   Command window is used for results

```
% Command Window
adaptive(@rty2,0,pi/2)
Gausss_quadrature(@rty2,0,pi/2,pi/6)
```

CHAPTER 6

Solution 1: (i) Construct an Mfile

```
% Mfile
function z=rst1(x,y)
z=-x+y;
end
```

This code along with the Euler's, Heun's and modified Euler's codes are used to calculate the results on the command window.

```
% Command Window
euler_Method(@rst1,1,0,0.01,1)
Heuns(@rst1,1,0,0.01,1)
Modified_euler(@rst1,1,0,0.01,1)
```

(ii) Construct an Mfile

```
% Mfile
function z=rst2(x,y)
z=y^4-(4/x)*y;
end
```

This code along with the Euler's, Heun's and modified Euler's codes are used to calculate the results on the command window.

```
% Command Window
euler_Method(@rst2,1,1,0.01,3)
Heuns(@rst2,1,1,0.01,3)
Modified_euler(@rst2,1,1,0.01,3)
```

(iii) Construct an Mfile

```
% Mfile
function z=rst3(x,y)
z=(x^2-x*y+y^2)/x*y;
end
```

This code along with the Euler's, Heun's and modified Euler's codes are used to calculate the results on the command window.

```
% Command Window
euler_Method(@rst3,2,1,0.01,2)
Heuns(@rst3,2,1,0.01,2)
Modified_euler(@rst3,2,1,0.01,2)
```

(iv) Construct an Mfile

```
% Mfile
function z=rst4(x,y)
z=2*(1-y)/x^2*sin(x);
end
```

This code along with the Euler's, Heun's and modified Euler's codes are used to calculate the results on the command window.

```
% Command Window
euler_Method(@rst4,2,1,0.01,2)
Heuns(@rst4,2,1,0.01,2)
Modified_euler(@rst4,2,1,0.01,2)
```

Solution 2: (i) Construct an Mfile

```
% Mfile
function z=rce1(x,y)
z=[y(2);2*y(2)-y(1)];
end
```

This code along with the Euler's, Heun's and modified Euler's codes for the systems of equations are used to calculate the results on the command window.

```
% Command Window
euler_Method_system(@rce1,[2 -1],0,0.1,1)
Heuns_Method_System(@rce1,[2 -1],0,0.1,1)
Modified_euler_Method_system(@rce1,[2 -1],0,0.1,1)
```

(ii) Construct an Mfile

```
% Mfile
function z=rke1(x,y)
```

```
z=[y(2);y(3);-4*y(3)-5*y(2)];
end
```

This code along with the Euler's, Heun's and modified Euler's codes for the systems of equations are used to calculate the results on the command window.

```
% Command Window
euler_Method_system(@rke1,[1 0 -1],0,0.1,1)
Heuns_Method_System(@rke1,[1 0 -1],0,0.1,1)
Modified_euler_Method_system(@rke1,[1 0 -1],0,0.1,1)
```

(iii) Construct an Mfile

```
% Mfile
function z=rke2(x,y)
z=[y(2);sin(x)-4*y(2)-y(1)];
end
```

This code along with the Euler's, Heun's and modified Euler's codes for the systems of equations are used to calculate the results on the command window.

```
% Command Window
euler_Method_system(@rke2,[1 0],0,0.1,1)
Heuns_Method_System(@rke2,[1 0],0,0.1,1)
Modified_euler_Method_system(@rke2,[1 0],0,0.1,1)
```

Solution 3: Construct an Mfile

```
% Mfile
function z=rce12(x,y)
l=1;
g=9.8;
Y=0.21;
K=2.1;
z=[y(2);(-g/l)*sin(y(1)+(k^2/l)*Y*cos(y(1)*sin(k*x)];
end
```

This code along with the Runge-Kutta code for the systems of equations are used to calculate the results on the command window.

Solution 4: (i) Construct an Mfile

```
% Mfile
function z=rc12(x,y)
z=[y(2);y(3);y(2)^2-y(1)*y(3)];
end
```

This code along with the Runge-Kutta code for the systems of equations are used to calculate the results on the command window.

(ii) Construct an Mfile

```
% Mfile
function z=rc13(x,y)
z=[y(2);-y(1)*y(2)^2)];
end
```

This code along with the Runge-Kutta code for the systems of equations are used to calculate the results on the command window.

(iii) Construct an Mfile

```
% Mfile
function z=rc13(x,y)
z=[y(2);y(3);-0.5*y(1)*y(3)];
end
```

This code along with the Runge-Kutta code for the systems of equations are used to calculate the results on the command window.

Solution 5: (i) Construct an Mfile

```
% Mfile
function z=rsp1(x,y)
z=y-x^2;
end
```

This code along with the Runge-Kutta-Fehlberg code for the systems of equations are used to calculate the results on the command window.

(ii) Construct an Mfile

```
% Mfile
function z=rsp2(x,y)
z=(x+2*x^3)*y^3-x*y
end
```

This code along with the Runge-Kutta-Fehlberg code for the systems of equations are used to calculate the results on the command window.

(iii) Construct an Mfile

```
% Mfile
function z=rsp3(x,y)
z=-(y+1)*(y+3)
end
```

This code along with the Runge-Kutta-Fehlberg code for the systems of equations are used to calculate the results on the command window.

Solution 6: (i) Construct an Mfile

```
% Mfile
function z=rp1(x,y)
z=[y(2);y(3);y(2)*log(y(3)-sin(y(1)))];
end
```

This code along with the Runge-Kutta-Fehlberg code for the systems of equations are used to calculate the results on the command window.

(ii) Construct an Mfile

```
% Mfile
function z=rp2(x,y)
z=[y(2);-(x^2-1)*y(2)-y(1)];
end
```

This code along with the Runge-Kutta-Fehlberg code for the systems of equations are used to calculate the results on the command window.

   (iii) Construct an Mfile

```
% Mfile
function z=rp3(x,y)
z=[y(2);-4*y(2)-y(1)+sin(y(1))];
end
```

This code along with the Runge-Kutta-Fehlberg code for the systems of equations are used to calculate the results on the command window.

Solution 7: Construct an Mfile

```
% Mfile
function z=rtp(x,y)
m=1.2;
f=21;
G=5;
% G=10;
z=(G/m)-(f/m)*y;
end
```

This code along with the predictor-corrector code is used to calculate the results on the command window.

Solution 8: (i) Construct an Mfile

```
% Mfile
function z=rq1(x,y)
z=[y(2);y(3);-0.2*(y(1)*y(3)+2*y(2)^2];
end
```

This code along with the predictor-corrector code for the systems of equations are used to calculate the results on the command window.

   (ii) Construct an Mfile

```
% Mfile
function z=rq2(x,y)
z=[y(2);-exp(y(1))*(1+y(2))];
end
```

This code along with the predictor-corrector code for the systems of equations are used to calculate the results on the command window.

CHAPTER 7

Solution 1: (i) Construct an Mfile

```
% Mfile
function z=awr1(x,y)
z=[y(2);(y(2)-x^5)/x];
end
```

This code along with the linear shooting code is used to calculate the results on the command window.

(ii) Construct an Mfile

```
% Mfile
function z=awr2(x,y)
z=[y(2);-9*y(1)];
end
```

This code along with the linear shooting code is used to calculate the results on the command window.

(iii) Construct an Mfile

```
% Mfile
function z=awr3(x,y)
z=[y(2);(-y(1)-x*y(2))/6*x^2];
end
```

This code along with the linear shooting code is used to calculate the results on the command window.

Solution 2: Construct an Mfile

```
% Mfile
function z=awts(x,y)
E=0.1;
l=0.1;
m=0.1;
T=0.1;
z=[y(2);(T/E*l)*y(1)+m*x*(x-5)/2*E*l];
end
```

This code along with the linear shooting code is used to calculate the results on the command window.

Solution 3: Construct an Mfile

```
% Mfile
function z=awst(x,y)
z=[y(2);y(1)+y(2);y(3);y(3)+y(4)];
end
```

This code along with the linear shooting code for the systems of equations is used to calculate the results on the command window.

Solution 4: Construct an Mfile

```
% Mfile
function z=awef(x,y)
z=[y(2);y(3);y(4);-(4/x)*y(1)^3];
end
```

Few changes will be made in the code mentioned in page 187. These files are used in the command window to get results.

Solution 5: Construct an Mfile

```
% Mfile
function z=awey(x,y)
z=[y(2);y(1)^2-y(2)];
end
```

This code along with the code mentioned in page 187 is used to calculate the results on the command window.

Solution 6:

```
% Mfile
function z=awey(x,y)
z=[y(2);y(3);y(2)^3-y(1)*y(3)];
end
% Mfile
function [z]=third_order_euler_secant(func,x0,h,xf)
% z is the output
% func, y1, x0, h and xf are the inputs
u1=1.3;
y1=[0 1 u1];
x=x0:h:xf; % Domain
y(:,1)=y1; % One column
for i=1:length(x)-1
  y(:,i+1)=y(:,i)+h*func(x(i),y(:,i)); % Euler's formula
end
t=y';
a1=t(end,2)-0;
u2=0.3;
y1=[0 1 u2];
x=x0:h:xf; % Domain
y(:,1)=y1; % One column
for i=1:length(x)-1
  y(:,i+1)=y(:,i)+h*func(x(i),y(:,i)); % Euler's formula
end
t=y';
a2=t(end,2)-0;
for numiter=1:5
  s=a1;
  r=a2;
  q=u1;
  u=u2;
  u3=u -((u-q)/(r-s))*(r); % Secant method
for i=1:length(x)-1
y1=[0 1 u3];
x=x0:h:xf; % Domain
y(:,1)=y1; % One column
  y(:,i+1)=y(:,i)+h*func(x(i),y(:,i)); % Euler's formula
end
t=y';
v=t(end,2)-0;
a1=r;
a2=v;
u1=u;
u2=u3;
end
z=t;
plot(x,z(:,2))
xlabel('x')
ylabel('y')
end
```

Solution 7: The procedure of this system is like solution 6 with the addition of one more second order equation.

Solution 8: Construct an Mfile

```
% Mfile
function z=awt(x,y)
z=[y(2);y(3);-2*y(1)*y(2)-y(1);y(5);-y(1)*y(5);y(7);-y(6)*y(5)];
end
```

The procedure of this system is like solution 6 with the addition of two more second order equations.

Solution 9: The procedure is like the code mentioned in page 196.

Solution 10: The procedure is like the code mentioned in page 196.

Solution 11: The procedure is like the code mentioned in page 196.

Solution 12: The procedure is like the code mentioned in page 196.

Solution 13: Apply the central difference formula we get

$$y_{i+1} + y_i\left(-2 - h^2\right) + y_{i-1} = h^2 t_i\left(t_i - 4\right).$$

Using the code mentioned in page 206, replace the diagonal entries with $-2 - h^2$ lower diagonal entries with 1 and upper diagonal entries with 1 (excluding first and last row) and set the boundary conditions in the first and last row.

Solution 14: Apply the central difference formula we get

$$y_{i+1} - 2y_i + y_{i-1} = h^2 exp\left(y_i\right).$$

Using the code mentioned in page 206, replace the diagonal entries with -2 lower diagonal entries with 1 and upper diagonal entries with 1 (excluding first and last row) and set the boundary conditions in the first and last row.

Solution 15: The procedure is like the code mentioned for the nonlinear equations.

Solution 16: Apply the central difference formula we get

$$\left(1 + 0.25h\right)f_{i+1} - 2f_i + \left(1 - 0.25h\right)f_{i-1} = 0,$$

$$g_{i+1} + g_i\left(-2 + h^2\right) + g_{i-1} = 0.$$

We have

```
% Mfile
function Block_Band
xx(1)=0;
n=3;
h=0.1;
for i=2:n
  xx(i)=xx(i-1)+h;
end
```

```
a{1}=[1 0;0 1];
c{1}=[0 0;0 0];
for j=2:n-1
  a{j}=[-2 0;0 -2+h^2];
end
a{n}=[1 0;0 1];
for j=2:n-1
  b{j}=[1-0.25*h 0;0 1];
end
b{n}=[0 0;0 0];
for j=2:n-1
  c{j}=[1+0.25*h 0;0 1];
end
c{n}=[0 0;0 0];
r1(1)=0;
  r2(1)=3;
for j=2:n-1
  r1(j)=0;
  r2(j)=0;
end
r1(n)=300;
  r2(n)=0;
gamma{1}=inv(a{1})*c{1};
for j=2:n
  a{j}=a{j}-(b{j}*gamma{j-1});
  gamma{j}=inv(a{j})*c{j};
end
for j=1:n
  rr{j}=[r1(j);r2(j)];
end
y{1}=inv(a{1})*rr{1};
for j=2:n
  y{j}=inv(a{j})*(rr{j}-b{j}*y{j-1});
end
x{n}=y{n};
for j=np-1:-1:1
  x{j}=y{j}-(gamma{j})*x{j+1};
end
for j=n:-1:1
  k(:,j)=x{j}(1,1);
  z(:,j)=x{j}(2,1);
end
k=k';
z=z';
```

References

1. Akai, T.J., *Applied Numerical Method for Engineers*, John Wiley and Sons, New York, 1993.
2. Ayyub, B.M., and R.H. Mccuen, *Applied Numerical Methods for Engineers*, Prentice-Hall, Upper Saddle River, NJ, 1996.
3. Boor, C.D.E., and S.D. Conte, *Elementary Numerical Analysis*, McGraw-Hill, New York, 1980.
4. Jaluria, Y., *Computer Methods for Engineering with MATLAB Applications*, Taylor and Francis Group, New York, 2012.
5. Burden, R.L., and J.D. Faires, *Numerical Analysis*, Fourth Edition, PWS-KENT Publishing Company, Boston, MA, 1988.
6. Chapra, S.C., *Applied Numerical Methods with MATLAB*, Second Edition, McGraw-Hill, New York, 1985.
7. Mathews, J.H., and K.D. Fink, *Numerical Methods Using MATLAB*, Fourth Edition, Pearson, 2004.
8. Fausett, V.L., *Applied Numerical Analysis Using MATLAB*, Second Edition, Pearson, 2008.
9. Jain, M.K., and S.K.R. Iyengar, and R.K. Jain, *Numerical Methods for scientific and Engineering Computation*, Sixth Edition, New Age International Publishers, Dehli North, 2012.
10. Bradie, B., *A Friendly Introduction to Numerical Anlaysis*, Pearson, Newport News, VA, 2007.
11. Hageman L.A., and D.M. Yound, *Applied Iterative Methods*, Academic Press, New York, 1981.
12. Hoffman, J.D., *Numerical Methods for Engineering and Scientists*, McGraw, New York, 1992.
13. Kharab, A., and R.B. Guenther, *An Introduction to Numerical Methods A MATLAB Approach*, Fourth Edition, Taylor and Francis, London, 2018.
14. Esfandiari, R.S., *Numerical Methods for Engineering and Scientists Using MATLAB*, Second Edition, Taylor and Francis, London, 2017.
15. Alefred, B.M., and J. Herzberger, *Introduction to Interval Computations*, Academic Press, New York, 1983.
16. Carnahan, B., H.A. Luther, and J.O. Wilkers, *Applied Numerical Methods*, John Willey & Sons, New York, 1969.
17. Birkhoff, G., and G.C. Rota, *Ordinary Differential Equations*, John Willey & Sons, New York, 1978.
18. Bracewell, R.N., *The Fourier Transform and Its Applications*, McGraw-Hill, New York, 1978.
19. Brent, R.P., *Algorithms for Minimization Without Derivatives*, Prentice-Hall, Englewood Cliffs, NJ, 1973.
20. Chapra, S.C., *Numerical Methods for Engineers: With Personal Computer Applications*, McGraw-Hill, New York, 1985.
21. Cheney W., and D. Kincaid, *Numerical Mathematics and Computing*, Fourth Edition, Brooks/Cole Publishing Group, New York, 1999.
22. Dahlquist, G., and A. Bjorck, *Numerical Methods*, Prentice-Hall, Englewood Cliffs, NJ, 1974.
23. Datta, B.N., *Numerical Linear Algebra and Applications*, Brooks/Cole, Pacific Grove, CA, 1995.
24. Davis, P.J., and P. Rabinowitz, *Methods for Numerical Integration*, Second Edition, Academic Press, New York, 1984.
25. Eljindi, M., and A. Kharab, The Quadratic Method for Computing the Eigenpairs of a Matrix, *IJCM*, 73, 530–534, 2000.
26. Evans, G., *Practical Numerical Analysis*, John Willey & Sons, Chichester, 1995.
27. Fausett, L.V., *Applied Numerical Analysis Using MATLAB*, Prentice-Hall, Upper Siddle River, NJ, 1999.
28. Forsythe, G.E., M.A. Malcolm, and C.B. Moler, *Computer Methods for Mathematical Computations*, Prentice-Hall, Englewood Cliffs, NJ, 1977.
29. Gerald, C.F., and P.O. Wheatley, *Applied Numerical Analysis*, Addison-Wesley, Reading, MA, 1989.
30. Gill, P.E., W. Murray, and M.H. Wright, *Numerical Linear Algebra and Optimization*, Addison-Wesley, Redwood City, CA, 1991.
31. Golub, G., and C. Van Loan, *Matrix Computations*, John Hopkins Press, Baltimore, 1983.
32. Golub, G., and J.M. Ortega, *Scientific Computing and Differential Equations: An Introduction to Numerical Methods*, Academic Press, Inc., Boston, MA, 1992.
33. Greenspan, D., and V. Casulli, *Numerical Analysis for Applied Mathematics, Science and Engineering*, Addison Wesley, New York, 1988.

34. Hager, W.W., *Applied Numerical Linear Algebra Methods*, Prentices-Hall, Englewood Cliffs, NJ, 1988.
35. Hanselman, D., and B. Littlefield, *Mastering MATLAB 5: A Comprehensive Tutorial and Reference*, Prentice-Hall, Upper Siddle River, NJ, 1998.
36. Heath, J.D., *cientific Computing: An Introductory Survey*, McGraw-Hill, New York, 1992.
37. Horn, R.A., and C.R. Johnson, *Matrix Analysis*, Cambridge University Press, Cambridge, 1985.
38. Hultquist, P.F., *Numerical Methods for Engineering and Computer Science*, Benjamin/Cummings Publishing Company, 1988.
39. Isaacson, E., and H.B. Keller, *Analysis of Numerical Methods*, Second Edition, Willey, New York, 1990.
40. Johnston, R.L., *Numerical Methods: A Software Approach*, John Willey & Sons, New York, 1982.
41. Kahaner, D., C. Moler, and S. Nash, *Numerical Methods and Software*, Prentice-Hall, Englewood Cliffs, NJ, 1989.
42. Kernighan, B.W., and R. Pike, *The Practice of Programming*, Addiseon-Wesley, Reading, MA, 1999.
43. Kings, J.T., *Numerical Methods for Ordinary Differential Systems*, John Wiley and Sons, Chichester, 1991.
44. Lambert, J.D., The Initial Value Problems for Ordinary Differential Equations, in *The State of Art in numerical Analysis*, D. Jacobs, Editor Academic Press, New York, 4, 156–160, 1977.
45. Lawson, C. L., and R.J. Hanson, *Solving Leclsi-Square Problems*, Prentice-Hall, Englewood Cliffs, NJ, 1974.
46. Lindfield, G.R., and J.E.T. Penny, *Microcomputers in Numerical Analysis*, Ellis Horwood, Chichester, 1989.
47. Marchand, P., *Graphics and GUIs with MATLAB*, Third Edition, CRC Press, Boca Raton, FL, 2008.
48. Maron, M.J., and R.J. Lopez, *Numerical Analysis: A Practical Approach*, Third Edition, Wadsworth, Belmont, CA, 1991.
49. Mathews, J.H., *Numerical Methods for Mathematics, Science and Engineering*, Second Edition, Prentice-Hall, Englewood Cliffs, NJ, 1992.
50. Mcneary, S.S., *Introduction to Computational Methods for Students of Calculus*, Prentices-Hall, Englewood Cliffs, NJ, 1973.
51. Miller, W., *The Engineering of Numerical Software*, Prentice-Hall, Englewood Cliffs, NJ, 1984.
52. Miller, W., *A Software Tools Sampler*, Prentice-Hall, Englewood Cliffs, NJ, 1987.
53. Morris, J.L., *Computational Methods in Elementary Theory and Applications of Numerical Analysis*, John Wiley & Sons, New York, 1983.
54. Ortega, J.M., and W.G. Poole, *An Introduction to Numerical Methods for Differential Equations*, Pitman Press, Marshfield, MA, 1981.
55. Polking, J.C., *Ordinary Differential Equations Using MATLAB*, Prentice-Hall, Englewood Cliffs, NJ, 1995.
56. Ramirez, R.W., *The FFT, Fundamentals and Concepts*, Prentice-Hall, Englewood Cliffs, NJ, 1985.
57. Rice, J.R., *Numerical Methods*, Software, and Analysis, IMSL Reference Edition, McGraw-Hill, New York, 1983.
58. Robinson, R.C., *An Introduction to Dynamical Systems*, American Mathematical Society, Evanston, IL, 2012.
59. Schiavone, P., C. Coutanda, and A. Mioduchawski, *Integral Methods in Science and Engineering*, Birkhauser, Bostan, 2002.
60. Schwartz, H.R., *Numerical Anlaysis: A Comprehensive Introduction*, Willey, New York, 1989.
61. Shampine, L.F., R.C Allen, and S. Pruess, *Fundamentals of Numerical Computing*, John Wiley & Sons, New York, 1997.
62. Shampine, L.F., and M.W. Reichelt, The MATLAB ODE Suite, *SIAM Journal on Numerical Computing*, 18, 1–22, 1997.
63. Shingareva, I., and C. Lizarraga-Celaya, *Solving Nonlinear Partial Differential Equations with Maple and Mathematica*, Springer, Wien, New York, 2011.
64. Stoer, J., and R. Bulirsch, *Introduction to Numerical Analysis*, Springer-Verlag, New York, 1993.
65. Tourin, A., *An Introduction to Finite Differential Methods for PDEs in Finance*, The Fields Institute, Toronto, ON, 2010.
66. Tricomi, F.G., and C.H.H. Baker, *Treatment of Integral Equations by Numerical Methods*, Birkhauser, Boston, MA, 2002.
67. Varga, R.S., *Matrix Iterative Anlaysis*, Prentics-Hall, Englewood Cliffs, NJ, 1962.
68. Brebbia, C.A., *The Boundary Element Method for Engineers*, Third Edition, McGraw-Hill, London, 1977.

69. Brent, R., Some Efficient Algorithms for Solving System of Non-Linear Equations, *SIAM Journal on Numerical Analysis*, 10, 327–344, 1973.
70. Bronson, R., and G.B. Costa, *Matrix Methods: Applied Linear Algebra*, Third Edition, Academic Press, New York, 2008.
71. Butcher, J.C., On Runge-Kutta processes of high Order, *Journal of the Australian Mathematical Society*, 4, 179–194, 1964.
72. Clocksin, W.F., and C.S. Mellish, *Programming in PROLOG: Using the ISO Standard*, Fifth Edition, Springer-Verlag, New York, 2004.
73. Collatz, L. *The Numerical Treatment of Differential Equations*, Third Edition, Springer-Verlag, Berlin, 1966.
74. Davis, P.J., and P. Rabinowitz, *Numerical Integration*, Ginn-Blaisdell, Waltham, MA, 1967.
75. Ferziger J., *Numerical Methods for Engineering Applications*, Second Edition, Wiley-Interscience, New York, 1988.
76. Forsythe, G., and W. Wasow, *Finite Difference Methods for Partial Differential Equations*, Wiley, New York, 1998.
77. Forsythe, G., and W. Wasow, *Finite Element Methods for Partial Differential Equations*, Wiley, New York, 1960.
78. Forsythe, G.E., M.A. Malcolm, and C.B. Moler, *Computers Methods for Mathematical Computations*, Prentice-Hall, Englewood Cliffs, NJ, 1977.
79. Fox, L., *Numerical Solutions of Ordinary and Partial Differential Equations*, Pergamon Press, Oxford, 1962.
80. Epperson, J. F., *An Introduction to Numerical Methods and Analysis*, John Wiley & Sons, New York, 2021.
81. Schäfer, M., *Computational Engineering: Introduction to Numerical Methods*, Springer, Berlin, 2006.
82. Holmes, M.H., *Introduction to Numerical Methods in Differential Equations*, Springer, New York, 2007.
83. Platen, E., *An Introduction to Numerical Methods for Stochastic Differential Equations*, *Acta Numerica*, 8, 197–246, 1999.

Index

@ command, 11

A

Adams closed/Adams-Moulton formulas, 162, 163
Adams multistep methods
    Adams closed/Adams-Moulton formulas, 162, 163
    Adams open/Adams-Bashforth formulas, 161, 162
    backward finite difference approximation, 161
    fourth order Runge-Kutta method, 163–164
    main steps for, 164–165
    predictor-corrector methods, 163, 165–168
    second order open Adams formula, 161
    Taylor series expansion, 161
    third order formula, 161
Adams open/Adams-Bashforth formulas, 161, 162
Adaptive quadrature, 122–124
Algebraic mapping, 65
Archimedes' principle, 89
Array functions
    det command, 7
    dot product, 6
    length, 6
    meshgrid, 6
    reshape, 6–7
    size, 5

B

Banded block tridiagonal matrix
    Crout's factorization method
        [A] = [L][U], 46
        lower diagonal entries, 47
        [L][Y] = [R], 47
        main steps for, 49–50
        upper diagonal entries, 47
        [U][X] = [Y], 47
        x1, x2, x3 values calculation, 49
    Doolittle's factorization method
        [A] = [L][U], 42
        lower diagonal entries, 42
        [L] [Y] = [R], 43
        main steps for, 45–46
        matrix multiplication, 42
        upper diagonal entries, 42
        [U][X] = [Y], 43
        x_1, x_2, x_3 values calculation, 44–45
Banded coefficient matrix
    Crout's factorization method
        L and U multiplication, 38
        lower diagonal entries, 38
        A = LU, 37
        LY = R, 38
        main steps for, 40–41
        upper diagonal entries, 38
        UX = Y, 39
        x_1, x_2, x_3 values calculation, 40

Doolittle's factorization method
    L and U multiplication, 34
    lower diagonal entries, 34
    A = LU, 33
    LY = R, 34
    main steps for, 36–37
    upper diagonal entries, 34
    UX = Y, 35
    x_1, x_2, x_3 values calculation, 36
    non-zero terms, 33
Basic commands, 1
Bisection method, 90–92, 94
Boundary value problems (BVPs)
    finite difference method
        for linear ordinary differential equations, 202–211
        for nonlinear ordinary differential equations, 211–221
    graphical behavior of thin rod, 175, 176
    shooting method
        fourth order Runge-Kutta method, 176
        graphical behavior, 175, 176
        initial value problems, 175, 176
        for linear ordinary differential systems, 177–182
        for nonlinear ordinary differential systems, 182–201
        numerical values of k's, 176, 177
        second order linear boundary value problem, 175
Break statement, 10
BVPs, *see* Boundary value problems (BVPs)

C

Cell arrays, 3–4
Choleski's factorization method, 31–33
Colon operator, 4
Conjugate gradient method, 56–60, 64
    $\nabla F = Ax - b$, 56
    initial guess x_0, 56
    $(alpha)_k$, 56
    $(beta)_k$ constant, 56, 57
    main steps for, 59–60
    s_k search direction, 56
Convergence criterion, 60–64
Cramer's rule, 15–18
Crout's factorization method, 208
    banded block tridiagonal matrix, 46–50
    banded tridiagonal matrix, 37–41
    LU factorization/decomposition, 28–31

D

Developing arrays, 3
Displaced water's volume (V_d), 89
Doolittle's factorization method
    banded block tridiagonal matrix, 42–46
    banded tridiagonal matrix, 33–37
    LU factorization/decomposition, 22–27

E

'else' condition, 9
'elseif' condition, 9
Euler's method
   error analysis, 133–135
   Euler's formula, 132
   general form, 132
   vs. Heun's method, 138
   for systems of ordinary differential equations, 135–137
Eye function, 5

F

Feval function, 11
Finite difference method, boundary value problems
     (BVPs)
   for linear ordinary differential equations
     central difference approximation, 202–203
     finite difference approximation, 202
     main steps for, 206
     second order equation, 203–205
     for system of linear ordinary differential equations,
      207–211
   for nonlinear ordinary differential equations
     boundary conditions, 212, 213
     central difference formula, 213
     known quantities, 212
     main steps for, 214–215
     for second order nonlinear ordinary differential
      equations, 211–215
     for system of nonlinear ordinary differential
      equations, 215–221
     tridiagonal system, 213
Fourth order Runge-Kutta method, 163–164
   commonly used version of, 146
   global truncation error, 146
   graphical illustration of, 146, 147
   local truncation error, 146
   main steps for, 147–148
   for systems of ordinary differential equations, 148–150
fprintf functions, 12–13

G

Gauss elimination method, 61
   augmented coefficient matrix, 18
   back substitution phase, 18, 19
   elimination phase, 18
   pivot elements, 19
   for systems of n equations, 19
Gaussian quadrature
   Gauss Legendre formula, 125, 126
   graphical behavior of, 125
   interpolating functions, 125
   main steps for, 127–128
   n-point formula error, 126
   points and weights, 126
   trapezoidal rule, 124, 125
   undetermined coefficients, 125
Gauss-Jacobi method, 51–53, 63
Gauss-Jordan elimination method, 20–22, 61
Gauss Legendre formula, 125, 126
Gauss-Seidel method, 54–56, 64

H

Heun's method
   vs. Euler's method, 138
   general form, 138
   local and global errors, 138
   main steps for, 139
   second order Runge-Kutta methods, 145
   for systems of ordinary differential equations, 139–141

I

'if' condition, 9
Initial value problems (IVPs)
   differential equation, 131
   multistep methods
     Adams multistep methods, 161–165
     graphical illustration of, 160, 161
     Milne's method, 169–170
   Newton's second law, 131
   single step methods
     Euler's method, 132–137
     graphical illustration of, 132
     Heun's method, 138–141
     modified Euler's method, 141–144
     Runge-Kutta methods (*see* Runge-Kutta methods)
Inline functions, 11–12
Intermediate value theorem, 90
IVPs, *see* Initial value problems (IVPs)

L

Lagrange's interpolation formula, 75–78
Linear algebraic equations
   banded block tridiagonal matrix
     Crout's factorization method, 46–50
     Doolittle's factorization method, 42–46
   banded coefficient matrices
     Crout's factorization method, 37–41
     Doolittle's factorization method, 33–37
     non-zero terms, 33
   conjugate gradient method, 56–60, 64
   convergence criterion, 60–64
   Cramer's rule, 15–18
   Gauss elimination method, 61
     augmented coefficient matrix, 18
     back substitution phase, 18, 19
     elimination phase, 18
     pivot elements, 19
     for systems of n equations, 19
   Gauss-Jacobi method, 51–53, 63
   Gauss-Jordan elimination method, 20–22, 61
   Gauss-Seidel method, 54–56, 64
   horizontal and vertical equilibrium at nodes, 15
   LU factorization/decomposition
     Choleski's factorization method, 31–33
     Crout's factorization method, 28–31
     Doolittle's factorization method, 22–27
   single triangular system, 15, 16
Linear ordinary differential equations, finite difference
     method
   central difference approximation, 202–203
   finite difference approximation, 202
   main steps for, 206

second order equation, 203–205
for system of linear ordinary differential equations, 207–211
Linear ordinary differential systems, shooting method
  initial value problem, 179
  main steps for, 177–178, 181–182
  numerical values of k's, 179, 180
Linspace operator, 4
LU factorization/decomposition
  Choleski's factorization method, 31–33
  Crout's factorization method
    $AX = B$, 28
    L and U, 28, 29
    $A = LU$, 28
    main steps for, 30–31
    y_1, y_2 and y_3 values calculation, 30
  Doolittle's factorization method
    $AX = B$, 23
    inputs, 23
    $A = LU$, 23
    $LUX = B$, 23
    $LY = B$, 24
    main steps for, 26–27
    upper diagonal entries, 24
    $UX = Y$, 24
    y_1, y_2 and y_3 values calculation, 25

M

Matrix manipulation, 1–3
M.file usage, 8–9
Milne's method, 169–170
Modified Euler's method
  approximate average slope, 141
  graphical illustration of, 141
  local and global errors, 141
  main steps for, 142
  second order Runge-Kutta methods, 145–146
  slope at midpoint calculation, 141
  for systems of ordinary differential equations, 143–144
  trapezoidal rule, 141

N

Neville's method, 78–80
Newton Raphson method
  convergence criteria, 97
  derivation of, 95
  disadvantage, 96
  equation of tangent line, 95
  graphical behavior of, 95
  main steps for, 98–99
  stopping criteria, 96
  for system of equations, 101–104
Newton Raphson method combined with Euler's method
  for coupled nonlinear ordinary differential equations
    initial conditions, 198
    initial guesses, 198
    initial value problem, 197
    main steps for, 200–201
    solution of, 199–200
  for second order nonlinear problems
    initial conditions, 194

    initial guess, 194
    initial value problem, 193
    main steps for, 196–197
    solution of, 194–196
Newton's backward interpolation formula, 70–72
Newton's divided difference interpolation
  coefficients, 73
  divided difference table, 73, 74
  error in, 73
  main steps for, 74–75
  second divided difference, 73
  third divided difference, 73
  value of f(9), 74
Newton's forward interpolation formula
  diagonal entries, 68
  differentiation operator, 68
  error calculation, 67
  forward difference operation, 67, 68
  main steps for, 69–70
Newton's second law, 131
Nonlinear ordinary differential equations, finite difference method
  boundary conditions, 212, 213
  central difference formula, 213
  known quantities, 212
  main steps for, 214–215
  for second order nonlinear ordinary differential equations, 211–215
  for system of nonlinear ordinary differential equations, 215–221
  tridiagonal system, 213
Nonlinear ordinary differential systems, shooting method
  graphical behavior of, 183
  initial value problem, 183
  Newton Raphson method combined with Euler's method
    for coupled nonlinear ordinary differential equations, 197–201
    for second order nonlinear problems, 193–197
  root finding methods, 183
  Secant method combined with Euler's method
    for coupled nonlinear ordinary differential equations, 188–193
    for second order nonlinear problems, 183–188
Numerical integration
  adaptive quadrature, 122–124
  Gaussian quadrature
    Gauss Legendre formula, 125, 126
    graphical behavior of, 125
    interpolating functions, 125
    main steps for, 127–128
    n-point formula error, 126
    points and weights, 126
    trapezoidal rule, 124, 125
    undetermined coefficients, 125
  Newton-Cotes formulas
    open and closed integration, 107, 108
    $Pn(x)$ polynomial, 107
    Simpson's 1/3 rule, 111–115
    Simpson's 3/8 rule, 115–117
    trapezoidal rule, 107–111
  Richardson extrapolation
    numerical method, 117

Romberg integration, 118–122
total error, 118
truncation error, 118

O

Ones function, 4

P

Polynomial interpolation
    algebraic mapping, 65
    cubic spline interpolation method
        cubic model, 80
        cubic polynomial, 80, 83
        main steps for, 83–85
        natural splines, 80, 82
        second derivative, 80
        values of coefficients C's, 82, 83
    errors in, 65–67
    first order interpolating polynomial, 65, 66
    Lagrange's interpolation formula, 75–78
    Neville's method, 78–80
    Newton's backward interpolation formula, 70–72
    Newton's divided difference interpolation
        coefficients, 73
        divided difference table, 73, 74
        error in, 73
        main steps for, 74–75
        second divided difference, 73
        third divided difference, 73
        value of f(9), 74
    Newton's forward interpolation formula
        diagonal entries, 68
        differentiation operator, 68
        error calculation, 67
        forward difference operation, 67, 68
        main steps for, 69–70
    population data table, 65
    second order interpolating polynomial, 65, 66
Predictor-corrector methods, 163, 165–168

R

Ralston's method, 146
RAND function, 5
Regula falsi method
    chord equation, 93
    graphical behavior of false position, 92, 93
    intersection of chord, 93
    main steps for, 94–95
    stopping criteria, 93
Return command, 10–11
Richardson extrapolation
    numerical method, 117
    Romberg integration, 118–122
    total error, 118
    truncation error, 118
Romberg integration
    general form of, 119
    main steps for, 120–122
    $O(h^4)$ approximation, 120
    $O(h^6)$ approximation, 120
    $O(h^8)$ approximation, 120

total error (TE), 118, 119
Root finding methods
    bisection method, 90–92
    Newton Raphson method, 95–99, 101–104
    regula falsi method, 92–95
    secant method, 99–101
Runge-Kutta-Fehlberg technique
    coefficients, 151
    description, 151
    local error, 151
    main steps for, 153–155
    numerical solutions of, 153
    for systems of ordinary differential equations, 155–160
    value of h, 151
    values of coefficients, 153
Runge-Kutta methods
    fourth order Runge-Kutta method
        commonly used version of, 146
        global truncation error, 146
        graphical illustration of, 146, 147
        local truncation error, 146
        main steps for, 147–148
        for systems of ordinary differential equations, 148–150
    general form, 144
    Runge-Kutta-Fehlberg technique
        coefficients, 151
        description, 151
        local error, 151
        main steps for, 153–155
        numerical solutions of, 153
        for systems of ordinary differential equations, 155–160
        value of h, 151
        values of coefficients, 153
    second order form of
        equations with constants, 145
        Heun method, 145
        infinite numbers of, 145
        modified Euler's method, 145–146
        Ralston's method, 146

S

Secant method, 99–101
Secant method combined with Euler's method
    for coupled nonlinear ordinary differential equations
        approximate solution of nonlinear boundary value problem, 188–191
        boundary value system, 188
        initial guesses, 188
        initial value problem, 188
        main steps for, 191–193
    for second order nonlinear problems
        approximate solution of nonlinear boundary value problem, 184–186
        general form of, 184
        initial guess, 184, 185
        initial value problem, 183, 184
        main steps for, 186–188
Second order open Adams formula, 161
Second order Runge-Kutta methods
    equations with constants, 145
    Heun method, 145

infinite numbers of, 145
modified Euler's method, 145–146
Ralston's method, 146
Shooting method, boundary value problems (BVPs)
fourth order Runge-Kutta method, 176
graphical behavior, 175, 176
initial value problems, 175, 176
for linear ordinary differential systems, 177–182
for nonlinear ordinary differential systems, 182–201
numerical values of k's, 176, 177
Simpson's 1/3 rule
area under the curve, 112
graphical illustration of, 112
integral $f(x)$ function, 111
main steps for, 114–115
truncation error, 113
Simpson's 3/8 rule, 115–117
Specific gravity of sphere material, 89

T

Target loop, 10

Third order open Adams formula, 161
Three-dimensional graph, 8
Trapezoidal rule
composite trapezoidal rule, 109
function f(x), 107
graphical illustration of, 107, 108
integral T, 108
main steps for, 110–111
step size, 107
truncation error, 109
Two-dimensional graph, 7–8

V

Volume of sphere (V_s), 89

Z

Zeros function, 4–5

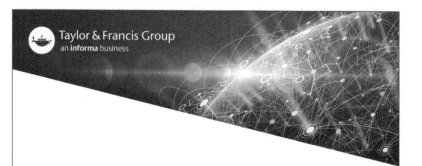